Introduction to Trigonometry

Introduction to Trigonometry

Isabella Hughes

Willford Press
www.willfordpress.com

Published by Willford Press,
118-35 Queens Blvd., Suite 400,
Forest Hills, NY 11375, USA

ISBN: 978-1-64728-349-0

Cataloging-in-Publication Data

Introduction to trigonometry / Isabella Hughes.
p. cm.
Includes bibliographical references and index.
ISBN 978-1-64728-349-0
1. Trigonometry. 2. Geometry. 3. Mathematics. I. Hughes, Isabella.
QA531 .I58 2022
516.24--dc23

For information on all Willford Press publications
visit our website at www.willfordpress.com

WILLFORD PRESS

Contents

Preface

The branch of mathematics which studies the relationship between side lengths and angles of triangles is known as trigonometry. This field of study is considered to be the foundation of all applied geometry. The real functions which relate an angle of a right-angled triangle to ratios of two side lengths are called trigonometric functions. They have a wide range of applications in astronomy, music theory, electronics, medical imaging and optics. Astronomy uses the trigonometric technique of triangulation to measure the distance to nearby stars. A few important concepts within trigonometry are triangle identities and trigonometric identities. This book presents the complex subject of trigonometry in the most comprehensible and easy to understand language. It will prove to be immensely beneficial to students and researchers in this field. Coherent flow of topics and extensive use of examples make this book an invaluable source of knowledge.

To facilitate a deeper understanding of the contents of this book a short introduction of every chapter is written below:

Chapter 1- The branch of mathematics that deals with the study of relationships between side lengths and angles of triangles. They can be defined using the unit circle and are used in the areas of geodesy, celestial mechanics, optics and acoustics, and navigation. This chapter has been carefully written to provide an easy understanding of trigonometry.

Chapter 2- Trigonometric functions are the angle functions which gives a relationship between an angle of a right-angles triangle and ratios of two side lengths. It includes sine, cosine, tangent, cosecant, secant and cotangent. This is an introductory chapter which will briefly introduce about these trigonometric functions.

Chapter 3- Inverse trigonometric functions are the inverse functions of sine, cosine, tangent, cotangent, secant, and cosecant functions. They can be used for obtaining an angle from any of the angle's trigonometric ratios. All these inverse trigonometric functions have been carefully analyzed in this chapter.

Chapter 4- The graphs of trigonometric functions of sine, cosine, tangent, cotangent, secant, and cosecant functions are periodic in nature. The graph cycle is repeated after every angle of 180 degrees. This chapter closely examines these graphs of trigonometric functions to provide an extensive understanding of the subject.

Chapter 5- Pythagorean identities, double angle formulas, half-angle formulas, triple-angle formulas, etc. are studied within trigonometric identities. Law of sines, cosines, tangents, Morrie's law, De Moivre's theorem, etc. fall under the domain of trigonometric laws. The topics elaborated in this chapter will help in gaining a better perspective about these trigonometric identities and laws.

Finally, I would like to thank the entire team involved in the inception of this book for their valuable time and contribution. This book would not have been possible without their efforts. I would also like to thank my friends and family for their constant support.

Isabella Hughes

1

Trigonometry: An Introduction

The branch of mathematics that deals with the study of relationships between side lengths and angles of triangles. They can be defined using the unit circle and are used in the areas of geodesy, celestial mechanics, optics and acoustics, and navigation. This chapter has been carefully written to provide an easy understanding of trigonometry.

Trigonometry is the branch of mathematics concerned with specific functions of angles and their application to calculations. There are six functions of an angle commonly used in trigonometry. Their names and abbreviations are sine (sin), cosine (cos), tangent (tan), cotangent (cot), secant (sec), and cosecant (csc). These six trigonometric functions in relation to a right triangle are displayed in the figure. For example, the triangle contains an angle A, and the ratio of the side opposite to A and the side opposite to the right angle (the hypotenuse) is called the sine of A, or sin A; the other trigonometry functions are defined similarly. These functions are properties of the angle A independent of the size of the triangle, and calculated values were tabulated for many angles before computers made trigonometry tables obsolete. Trigonometric functions are used in obtaining unknown angles and distances from known or measured angles in geometric figures.

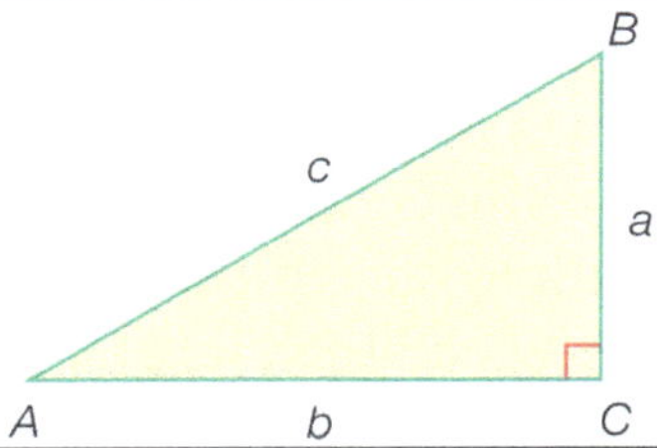

$$\sin A = \frac{a}{c} = \frac{\text{side opposite}}{\text{hypotenuse}} \qquad \csc A = \frac{c}{a} = \frac{\text{hypotenuse}}{\text{side opposite}}$$

$$\cos A = \frac{b}{c} = \frac{\text{side adjacent}}{\text{hypotenuse}} \qquad \sec A = \frac{c}{b} = \frac{\text{hypotenuse}}{\text{side adjacent}}$$

$$\tan A = \frac{a}{b} = \frac{\text{side opposite}}{\text{side adjacent}} \qquad \cot A = \frac{b}{a} = \frac{\text{side adjacent}}{\text{side opposite}}$$

The six trigonometric functions: Based on the definitions, various simple relationships exist among the functions. For example, csc A = 1/sin A, sec A = 1/cos A, cot A = 1/tan A, and tan A = sin A/cos A.

Trigonometry developed from a need to compute angles and distances in such fields as astronomy, mapmaking, surveying, and artillery range finding. Problems involving angles and distances in one plane are covered in plane trigonometry. Applications to similar problems in more than one plane of three-dimensional space are considered in spherical trigonometry.

Principles of Trigonometry

Trigonometric Functions

A somewhat more general concept of angle is required for trigonometry than for geometry. An angle A with vertex at V, the initial side of which is VP and the terminal side of which is VQ, is indicated in the figure by the solid circular arc. This angle is generated by the continuous counterclockwise rotation of a line segment about the point V from the position VP to the position VQ. A second angle A′ with the same initial and termed as the coordinate axesinal sides, indicated in the figure by the broken circular arc, is generated by the clockwise rotation of the line segment from the position VP to the position VQ. Angles are considered positive when generated by counterclockwise rotations, negative when generated by clockwise rotations. The positive angle A and the negative angle A′ in the figure are generated by less than one complete rotation of the line segment about the point V. All other positive and negative angles with the same initial and terminal sides are obtained by rotating the line segment one or more complete turns before coming to rest at VQ.

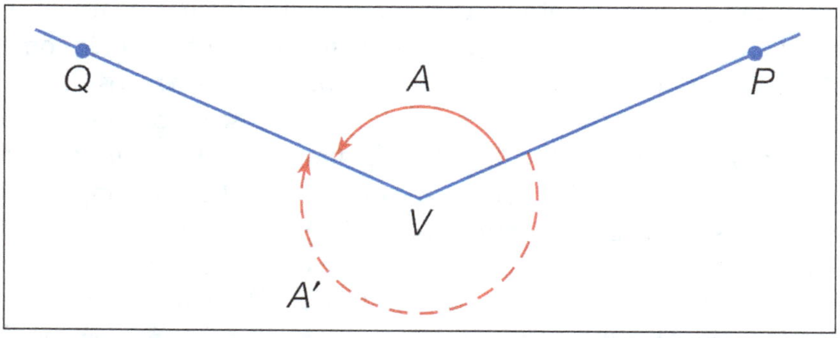

General angle: This figure shows a positive general angle A, as well as a negative general angle A'.

Numerical values can be assigned to angles by selecting a unit of measure. The most common units are the degree and the radian. There are 360° in a complete revolution, with each degree further divided into 60′ (minutes) and each minute divided into 60″ (seconds). In theoretical work, the radian is the most convenient unit. It is the angle at the centre of a circle that intercepts an arc equal in length to the radius; simply put, there are 2π radians in one complete revolution. From these definitions, it follows that $1° = {}^{\pi}/_{180}$ radians.

Equal angles are angles with the same measure; i.e., they have the same sign and the same number of degrees. Any angle −A has the same number of degrees as A but is of opposite sign. Its measure, therefore, is the negative of the measure of A. If two angles, A and B, have the initial sides VP and VQ and the terminal sides VQ and VR, respectively, then the angle A + B has the initial and terminal sides VP and VR. The angle A + B is called the sum of the angles A and B, and its relation to A and B when A is positive and B is positive or negative is illustrated in the figure. The sum A + B is the angle the measure of which is the algebraic sum of the measures of A and B. The difference A − B is the sum of A and −B. Thus, all angles coterminal with angle A (i.e., with the same initial and terminal sides as angle A) are given by A ± 360n, in which 360n is an angle of n complete revolutions. The angles (180 − A) and (90 − A) are the supplement and complement of angle A, respectively.

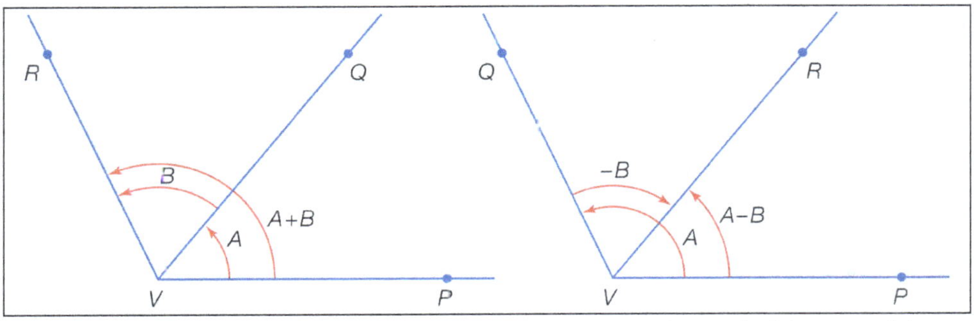

Addition of angles: The figure indicates how to add a positive or negative angle (B) to a positive angle (A).

Trigonometric Functions of an Angle

To define trigonometric functions for any angle A, the angle is placed in position on a rectangular coordinate system with the vertex of A at the origin and the initial side of A along the positive x-axis; r (positive) is the distance from V to any point Q on the terminal side of A, and (x, y) are the rectangular coordinates of Q.

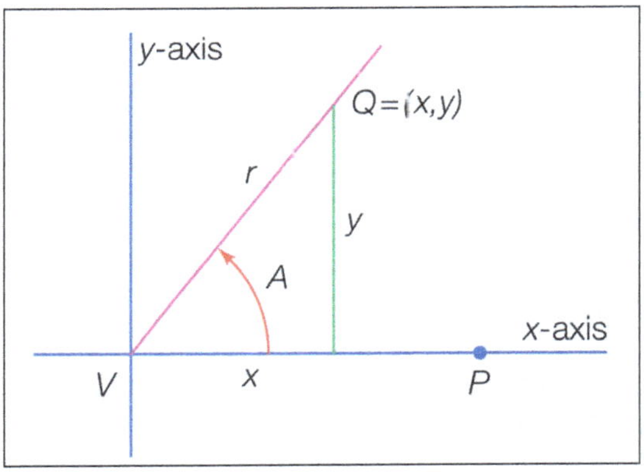

Angle in standard position: The figure shows an angle A in standard position, that is, with initial side on the x-axis.

The six functions of A are then defined by six ratios exactly as in the earlier case for the triangle given in the introduction. Because division by zero is not allowed, the tangent and secant are not defined for angles the terminal side of which falls on the y-axis, and the cotangent and cosecant are undefined for angles the terminal side of which falls on the x-axis. When the Pythagorean equality $x^2 + y^2 = r^2$ is divided in turn by r^2, x^2, and y^2, the three squared relations relating cosine and sine, tangent and secant, cotangent and cosecant are obtained.

Table: Negative Angles.

sin(-A)=-sin A	csc(-A) = -csc A
cos(-A) = cos A	sec(-A) = sec A
tan(-A) = -tan A	cos(-A) = -cot A

If the point Q on the terminal side of angle A in standard position has coordinates (x, y), this point will have coordinates (x, −y) when on the terminal side of −A in standard position. From this fact and the definitions are obtained further identities for negative angles. These relations may also be stated briefly by saying that cosine and secant are even functions (symmetrical about the y-axis), while the other four are odd functions (symmetrical about the origin).

It is evident that a trigonometric function has the same value for all coterminal angles. When n is an integer, therefore, sin (A ± 360n) = sin A; there are similar relations for the other five functions. These results may be expressed by saying that the trigonometric functions are periodic and have a period of 360° or 180°.

Table: Complementary angles and cofunctions.

$\sin(A \pm 90°) = \pm \cos A$	$\csc(A \pm 90°) = \pm \sec A$
$\cos(A \pm 90°) = \mp \sin A$	$\sec(A \pm 90°) = \mp \csc A$
$\tan(A \pm 90°) = -\cot A$	$\cot(A \pm 90°) = -\tan A$

When Q on the terminal side of A in standard position has coordinates (x, y), it has coordinates (−y, x) and (y, −x) on the terminal side of A + 90 and A − 90 in standard position, respectively. Consequently, six formulas equate a function of the complement of A to the corresponding cofunction of A.

Tables of Natural Functions

To be of practical use, the values of the trigonometric functions must be readily available for any given angle. Various trigonometric identities show that the values of the functions for all angles can readily be found from the values for angles from 0° to 45°. For this reason, it is sufficient to list in a table the values of sine, cosine, and tangent for all angles from 0° to 45° that are integral multiples of some convenient unit (commonly 1′). Before computers rendered them obsolete in the late 20th century, such trigonometry tables were helpful to astronomers, surveyors, and engineers.

Table: Common angles for trigonometry functions.

	0°	30°	45°	60°	90°
sin	0	1/2	$\sqrt{2}/2$	$\sqrt{3}/2$	1
cos	1	$\sqrt{3}/2$	$\sqrt{2}/2$	1/2	0
tan	0	$\sqrt{3}/3$	1	$\sqrt{3}$	undefined

For angles that are not integral multiples of the unit, the values of the functions may be interpolated. Because the values of the functions are in general irrational numbers, they are entered in the table as decimals, rounded off at some convenient place. For most purposes, four or five decimal places are sufficient, and tables of this accuracy are common. Simple geometrical facts alone, however, suffice to determine the values of the trigonometric functions for the angles 0°, 30°, 45°, 60°, and 90°. These values are listed in a table for the sine, cosine, and tangent functions.

Plane Trigonometry

In many applications of trigonometry the essential problem is the solution of triangles. If enough sides and angles are known, the remaining sides and angles as well as the area can be calculated, and the triangle is then said to be solved. Triangles can be solved by the law of sines and the law of cosines. To secure symmetry in the writing of these laws, the angles of the triangle are lettered A, B, and C and the lengths of the sides opposite the angles are lettered a, b, and c, respectively.

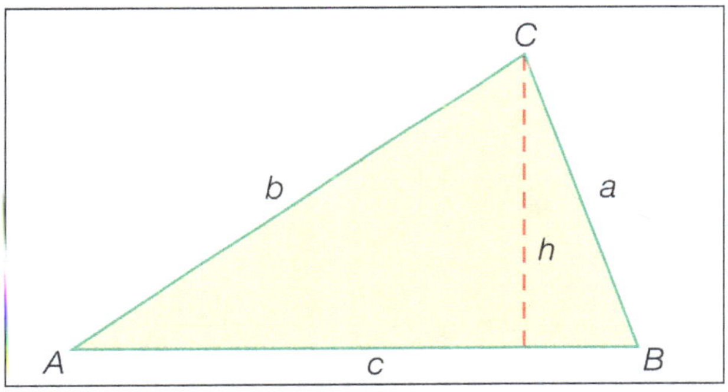

Standard lettering of a triangle.

In addition to the angles (A, B, C) and sides (a, b, c), one of the three heights of the triangle (h) is included by drawing the line segment from one of the triangle's vertices (in this case C) that is perpendicular to the opposite side of the triangle.

The law of sines is expressed as an equality involving three sine functions while the law of cosines is an identification of the cosine with an algebraic expression formed from the lengths of sides opposite the corresponding angles. To solve a triangle, all the known values are substituted into equations expressing the laws of sines and cosines, and the equations are solved for the unknown quantities. For example, the law of sines is employed when two angles and a side are known or when two sides and an angle opposite one are known. Similarly, the law of cosines is appropriate when two sides and an included angle are known or three sides are known.

Spherical Trigonometry

Spherical trigonometry involves the study of spherical triangles, which are formed by the intersection of three great circle arcs on the surface of a sphere. Spherical triangles were subject to intense study from antiquity because of their usefulness in navigation, cartography, and astronomy.

The angles of a spherical triangle are defined by the angle of intersection of the corresponding tangent lines to each vertex. The sum of the angles of a spherical triangle is always greater than the sum of the angles in a planar triangle (π radians, equivalent to two right angles). The amount by which each spherical triangle exceeds two right angles (in radians) is known as its spherical excess. The area of a spherical triangle is given by the product of its spherical excess E and the square of the radius r of the sphere it resides on—in symbols, Er^2.

Table: Common spherical trigonometry formulas.

Law of sines:	$\frac{\sin a}{\sin A}=\frac{\sin b}{\sin B}=\frac{\sin c}{\sin C}$
Law of cosines:	$\cos a=\cos b\cos c+\sin b\sin c\cos A$
	$\cos b=\cos a\cos c+\sin a\sin c\cos B$
	$\cos c=\cos a\cos b+\sin a\sin b\cos C$
Half-angle formulas:	$\tan\left(\frac{A}{2}\right)=\sqrt{\frac{\sin(s-b)\sin(s-c)}{\sin s\sin(s-a)}}$
	$\tan\left(\frac{B}{2}\right)=\sqrt{\frac{\sin(s-c)\sin(s-a)}{\sin s\sin(s-b)}}$
	$\tan\left(\frac{C}{2}\right)=\sqrt{\frac{\sin(s-a)\sin(s-b)}{\sin s\sin(s-c)}}$, where $s=\frac{a+b+c}{2}$
Half-side formulas:	$\tan\left(\frac{a}{2}\right)=\sqrt{\frac{-\cos S\cos(S-A)}{\cos(S-B)\cos(S-C)}}$
	$\tan\left(\frac{b}{2}\right)=\sqrt{\frac{-\cos S\cos(S-B)}{\cos(S-A)\cos(S-C)}}$
	$\tan\left(\frac{c}{2}\right)=\sqrt{\frac{-\cos S\cos(S-C)}{\cos(S-A)\cos(S-B)}}$, where $S=\frac{A+B+C}{2}$

By connecting the vertices of a spherical triangle with the centre O of the sphere that it resides on, a special "angle" known as a trihedral angle is formed. The central angles (also known as dihedral angles) between each pair of line segments OA, OB, and OC are labeled α, β, and γ to correspond to the sides (arcs) of the spherical triangle labeled a, b, and c, respectively. Because a trigonometric function of a central angle and its corresponding arc have the same value, spherical trigonometry formulas are given in terms of the spherical angles A, B, and C and, interchangeably, in terms of the arcs a, b, and c and the dihedral angles α, β, and γ. Furthermore, most formulas from plane trigonometry have an analogous representation in spherical trigonometry. For example, there is a spherical law of sines and a spherical law of cosines.

As was described for a plane triangle, the known values involving a spherical triangle are substituted in the analogous spherical trigonometry formulas, such as the laws of sines and cosines, and the resulting equations are then solved for the unknown quantities.

Table: Napier's analogies.

$\tan\left(\frac{a}{2}\right)\cos\left(\frac{B-C}{2}\right)=\tan\left(\frac{b+c}{2}\right)\cos\left(\frac{B+C}{2}\right)$
$\tan\left(\frac{a}{2}\right)\sin\left(\frac{B-C}{2}\right)=\tan\left(\frac{b-c}{2}\right)\sin\left(\frac{B+C}{2}\right)$
$\cot\left(\frac{A}{2}\right)\cos\left(\frac{b-C}{2}\right)=\tan\left(\frac{B+C}{2}\right)\cos\left(\frac{b+c}{2}\right)$
$\cot\left(\frac{A}{2}\right)\sin\left(\frac{b-C}{2}\right)=\tan\left(\frac{B-C}{2}\right)\cos\left(\frac{b+c}{2}\right)$

Many other relations exist between the sides and angles of a spherical triangle. Worth mentioning are Napier's analogies (derivable from the spherical trigonometry half-angle or half-side formulas), which are particularly well suited for use with logarithmic tables.

Analytic Trigonometry

Analytic trigonometry combines the use of a coordinate system, such as the Cartesian coordinate system used in analytic geometry, with algebraic manipulation of the various trigonometry functions to obtain formulas useful for scientific and engineering applications.

Trigonometric functions of a real variable x are defined by means of the trigonometric functions of an angle. For example, sin x in which x is a real number is defined to have the value of the sine of the angle containing x radians. Similar definitions are made for the other five trigonometric functions of the real variable x. These functions satisfy the previously noted trigonometric relations with A, B, 90°, and 360° replaced by x, y, $^{\pi}/_{2}$ radians, and 2π radians, respectively. The minimum period of tan x and cot x is π, and of the other four functions it is 2π.

In calculus it is shown that sin x and cos x are sums of power series. These series may be used to compute the sine and cosine of any angle. For example, to compute the sine of 10°, it is necessary to find the value of sin $\pi/_{18}$ because 10° is the angle containing $\pi/_{18}$ radians. When $\pi/_{18}$ is substituted in the series for sin x, it is found that the first two terms give 0.17365, which is correct to five decimal places for the sine of 10°. By taking enough terms of the series, any number of decimal places can be correctly obtained. Tables of the functions may be used to sketch the graphs of the functions.

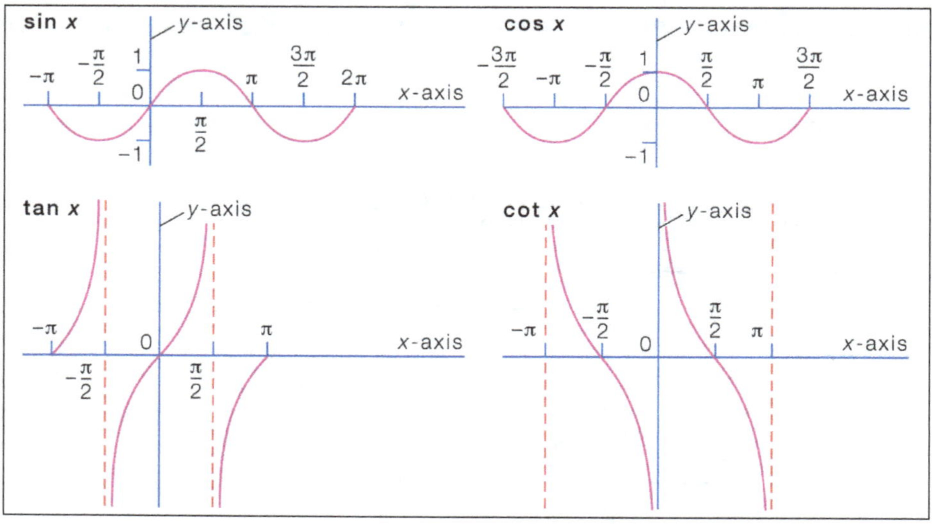

Graphs of some trigonometric functions. Each of these functions is periodic. Thus, the sine and cosine functions repeat every 2π, and the tangent and cotangent functions repeat every π.

Each trigonometric function has an inverse function, that is, a function that "undoes" the original function. For example, the inverse function for the sine function is written arcsin or $\sin^{-1}$, thus $\sin^{-1}(\sin x) = \sin(\sin^{-1} x) = x$. The other trigonometric inverse functions are defined similarly.

Coordinates and Transformation of Coordinates

Polar Coordinates

For problems involving directions from a fixed origin (or pole) O, it is often convenient to specify a point P by its polar coordinates (r, θ), in which r is the distance OP and θ is the angle that the direction of r makes with a given initial line. The initial line may be identified with the x-axis of rectangular Cartesian coordinates, as shown in the figure. The point (r, θ) is the same as (r, θ + 2nπ) for any integer n. It is sometimes desirable to allow r to be negative, so that (r, θ) is the same as (−r, θ + π).

Given the Cartesian equation for equation a curve, the polar equation for the same curve can be obtained in terms of the radius r and the angle θ by substituting r cos θ and r sin θ for x and y, respectively. For example, the circle $x^2 + y^2 = a^2$ has the polar $(r \cos \theta)^2 + (r \sin \theta)^2 = a^2$, which reduces to r = a. (The positive value of r is sufficient, if θ takes all values from $-\pi$ to π or from 0 to 2π). Thus the polar equation of a circle simply expresses the fact that the curve is independent of θ and has constant radius. In a similar manner, the line $y = x \tan \varphi$ has the polar equation $\sin \theta = \cos \theta \tan \varphi$, which reduces to $\theta = \varphi$. (The other solution, $\theta = \varphi + \pi$, can be discarded if r is allowed to take negative values).

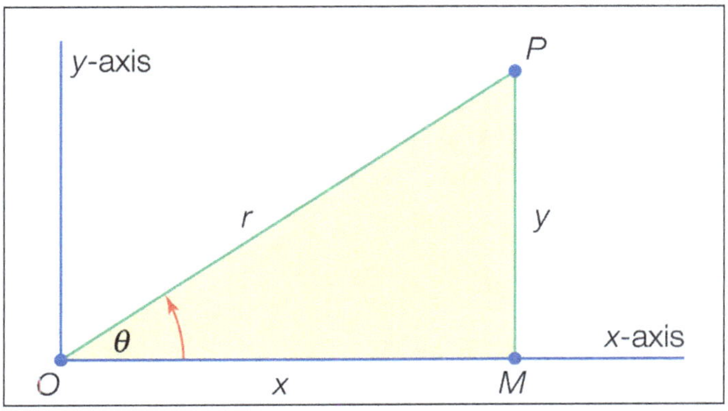

Cartesian and polar coordinates: The point labeled P in the figure resides in the plane. Therefore, it requires two dimensions to fix its location, either in Cartesian coordinates (x, y) or in polar coordinates (r, θ).

Transformation of Coordinates

A transformation of coordinates in a plane is a change from one coordinate system to another. Thus, a point in the plane will have two sets of coordinates giving its position with respect to the two coordinate systems used, and a transformation will express the relationship between the coordinate systems. For example, the transformation between polar and Cartesian coordinates is given by $x = r \cos \theta$ and $y = r \sin \theta$. Similarly, it is possible to accomplish transformations between rectangular and oblique coordinates.

In a translation of Cartesian coordinate axes, a transformation is made between two sets of axes that are parallel to each other but have their origins at different positions. If a point P has coordinates (x, y) in one system, its coordinates in the second system are given by $(x - h, y - k)$ where (h, k) is the origin of the second system in terms of the first coordinate system. Thus, the transformation of P between the first system (x, y) and the second system (x′, y′) is given by the equations $x = x' + h$ and $y = y' + k$. The common use of translations of axes is to simplify the equations of curves. For example, the equation $2x^2 + y^2 - 12x - 2y + 17 = 0$ can be simplified with the translations $x' = x - 3$ and $y' = y - 1$ to an equation involving only squares of the variables and a constant term: $(x')^2 + (y')^2/_2 = 1$. In other words, the curve represents an ellipse with its centre at the point (3, 1) in the original coordinate system.

A rotation of coordinate axes is one in which a pair of axes giving the coordinates of a point (x, y) rotate through an angle φ to give a new pair of axes in which the point has coordinates (x′, y′), as shown in the figure. The transformation equations for such a rotation are given by $x = x' \cos \varphi - y' \sin \varphi$ and $y = x' \sin \varphi + y' \cos \varphi$. The application of these formulas with $\varphi = 45°$ to the difference of squares, $x^2 - y^2 = a^2$, leads to the equation $x'y' = c$ (where c is a constant that depends on the value of a). This equation gives the form of the rectangular hyperbola when its asymptotes (the lines that a curve approaches without ever quite meeting) are used as the coordinate axes.

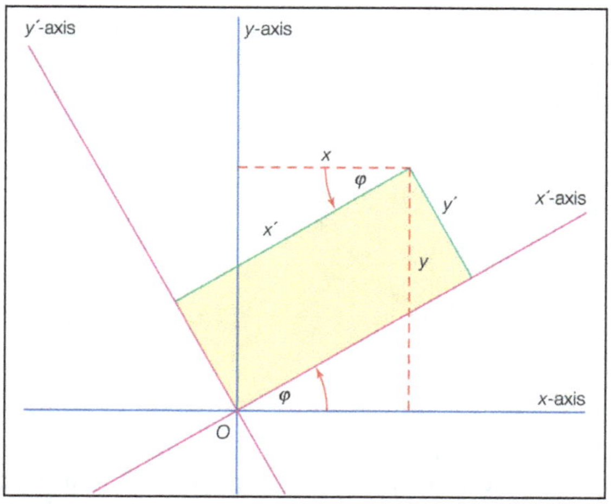

Rotation of axes: Rotating the coordinate axes through an angle φ changes the coordinates of a point from (x, y) to (x', y').

Angle

An angle is the rotation of a ray from an initial point to a terminal point. Some commonly used terms in angles are:

- Initial side: The original ray.
- Terminal side: The final position of the ray after rotation.
- Vertex: Point of rotation.
- Positive angle: If the direction of rotation is anticlockwise.
- Negative angle: If the direction of rotation is clockwise.

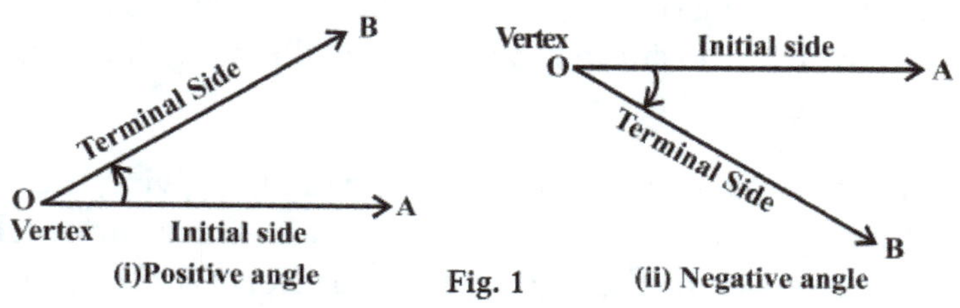

Fig. 1

Further, angle measurement is the amount of rotation performed by the initial side to get to the terminal side. There are several units for measuring angles.

Degree Measure

A complete revolution, i.e. when the initial and terminal sides are in the same position after rotating clockwise or anticlockwise, is divided into 360 units called degrees. So, if the rotation from the initial side to the terminal side is $\left(\frac{1}{360}\right)$th of a revolution, then the angle is said to have a measure of one degree. It is denoted as 1°.

We measure time in hours, minutes, and seconds, where 1 hour = 60 minutes and 1 minute = 60 seconds. Similarly, while measuring angles,

- 1 degree = 60 minutes denoted as $1° = 60'$.
- 1 minute = 60 seconds denoted as $1' = 60''$.

Here are some additional examples of angles with their measurements:

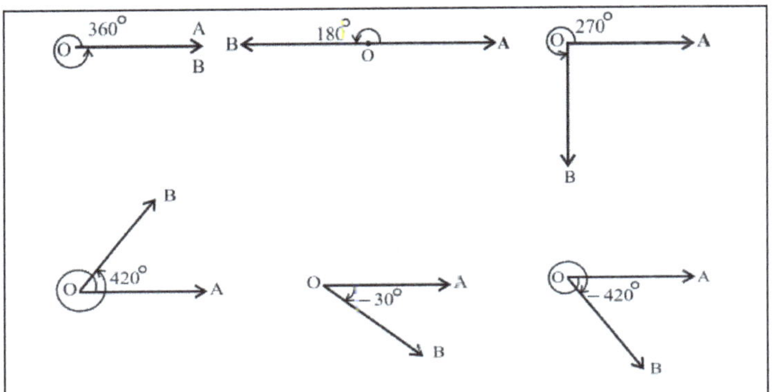

Fig. 2

Radian Measure

Radian measure is slightly more complex than the degree measure. Imagine a circle with a radius of 1 unit. Next, imagine an arc of the circle having a length of 1 unit. The angle subtended by this arc at the centre of the circle has a measure of 1 radian. Here is how it looks:

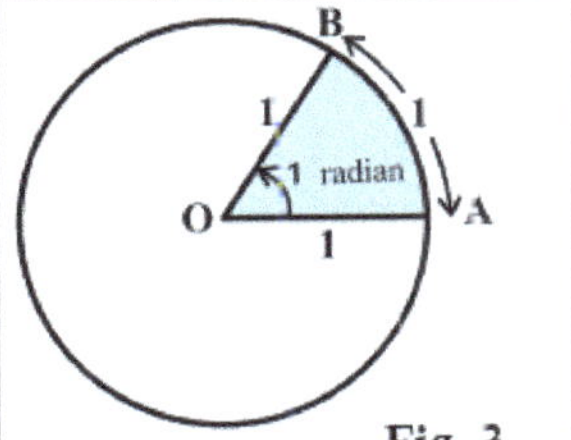

Fig. 3

Here are some more examples of angles that measure -1 radian, $1\frac{1}{2}$ radian, and $-1\frac{1}{2}$ radian.

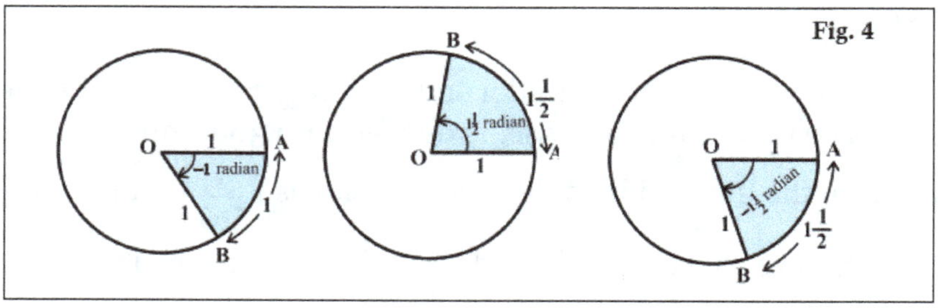

Fig. 4

The circumference of a circle = 2πr where r is the radius of the circle. Hence, for a circle with a radius of 1 unit, the circumference is 2π. Hence, one complete revolution of the initial side subtends an angle of 2π radian at the centre. Generalizing this, we have, In a circle of radius r, an arc of length r subtends an angle of 1 radian at the centre. Hence, in a circle of radius r, an arc of length l will subtend an angle = l/r radian. Generalizing this, we have, in a circle of radius r, if an arc of length l subtends an angle θ radian at the centre, then:

$$\theta = \frac{l}{r}$$
$$\Rightarrow l = r\theta.$$

The Relation between Radian and Real Numbers

Radian measures and real numbers are one and the same. Let's see how: consider a unit circle with centre O. Let A be any point on the circle and OA be the initial side of the angle as shown below:

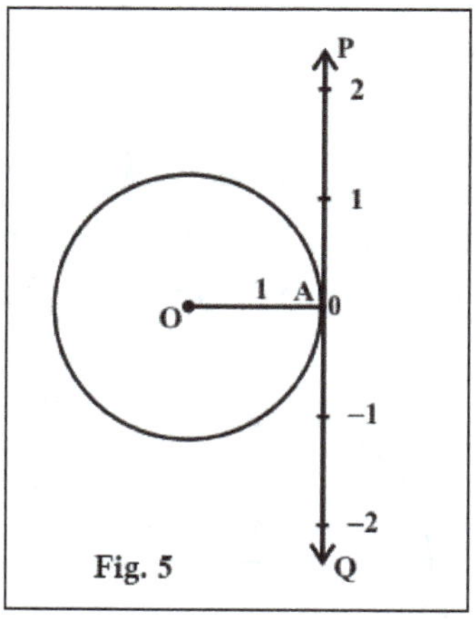

Fig. 5

Now, consider a line PAQ drawn tangential to the circle at point A. Also, let A be the real number zero. Hence, line AP represents the positive real numbers and line AQ represents the negative real numbers. Further, let's drag the line AP along the circumference of the circle in the anticlockwise direction.

Also, let's drag the line AQ along the circumference of the circle in the clockwise direction. After doing so, we can see that every real number corresponds to a radian measure and conversely.

The Relation between Degree and Radian Measures

By the definitions of degree and radian measures, we know that the angle subtended by a circle at the centre is:

- 360° – according to degree measure.
- 2π radian – according to radian measure.

Hence, 2π radian = 360° ⇒ π radian = 180°. Now, we substitute the approximate value of π as $\frac{22}{7}$ in the equation above and get, 1 radian $=\frac{180°}{\pi}=57°16'$ approximately. Also, $1°=\frac{\pi}{180°}$ radian = 0.01746 radian approximately. Further, here is a table depicting the relationship between degree and radian measures of some common angles:

Degree	30°	45°	60°	90°	180°	270°	360°
Radian	$\frac{\pi}{6}$	$\frac{\pi}{4}$	$\frac{\pi}{3}$	$\frac{\pi}{2}$	π	$\frac{3\pi}{2}$	2π

Notational Convention

Since degree and radian measures are the two most commonly used units in angle measurement, there is a convention in place for writing them.

- If you write angle θ°, then it means an angle whose degree measure is θ.
- If you write angle β, then it means an angle whose radian measure is β.

Also, note that the term 'radian' is usually omitted while writing the radian measure. Hence, π radian = 180° is simply written as π = 180°. Further, summing up the relationship between degree and radian measures, we have:

- Radian measure = $\frac{\pi}{180°}$ x Degree measure.
- Degree measure = $\frac{180°}{\pi}$ x Radian measure.

Example: Convert 40° 20′ into radian measure.

Solution: We know that 1° = 60′. Therefore, $20' = \frac{1}{3}$ degree. Hence, $40°20' = 40\frac{1}{3}$ degree $= \frac{121}{3}$ degree.

Also, we know that, Radian measure $= \frac{\pi}{180°} X$ Degree measure.

Therefore, the radian measure of $40°20' = \frac{\pi}{180} X \frac{121}{3} = \frac{121\pi}{540}$ radian.

Example: A wheel makes 360 revolutions in one minute. Through how many radians does it turn in one second?

Solution: We know that the wheel makes 360 revolutions in one minute. Hence, in one second it will make, $\frac{360}{60} = 6$ revolutions.

Also, we know that in one complete revolution, the wheel rotates through an angle of a 2π radian. Therefore, in 6 revolutions it will turn an angle of 6 x 2π radian = 12π radian. Hence, the wheel turns an angle of 12π radian in one second.

Unit Circle

The unit circle is a circle of radius 1 unit that is centered on the origin of the coordinate plane.

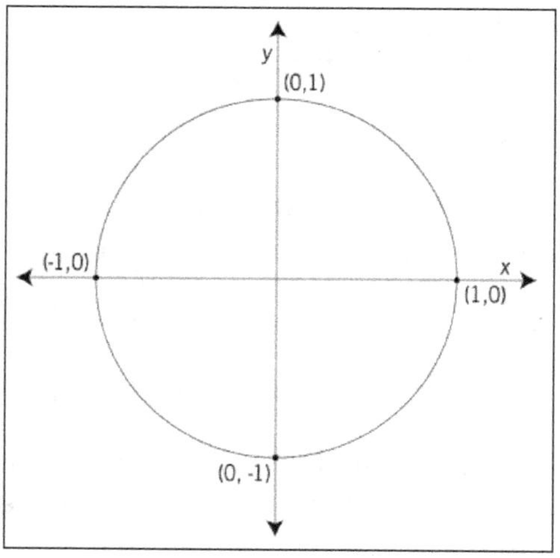

The unit circle is fundamentally related to concepts in trigonometry. The trigonometric functions can be defined in terms of the unit circle, and in doing so, the domain of these functions is extended to all real numbers.

The unit circle is also related to complex numbers. A unit circle can be graphed in the complex plane, and all roots of unity will lie on this circle.

Relation to Right Triangles

Every point on the unit circle corresponds to a right triangle with vertices at the origin and the point on the unit circle. The right triangle has leg lengths that are equal to the absolute values of the x and y coordinates, respectively.

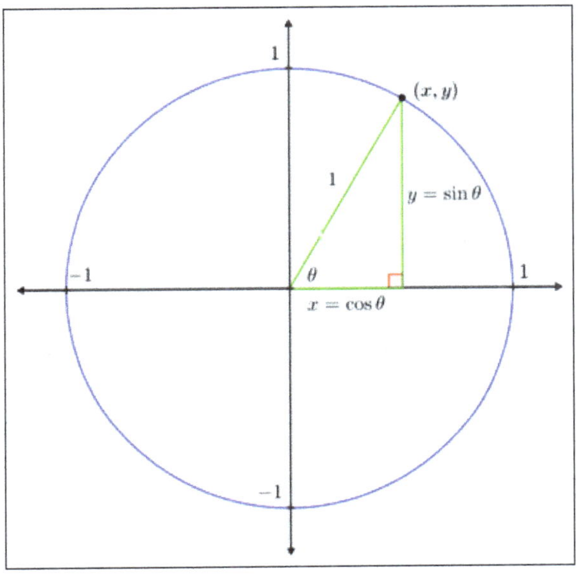

This right triangle is used to apply trigonometric relations.

$$\sin(\theta) = \frac{\text{opposite}}{\text{hypotenuse}} = \frac{b}{c}$$

$$\cos(\theta) = \frac{\text{adjacent}}{\text{hypotenuse}} = \frac{a}{c}$$

$$\tan(\theta) = \frac{\text{opposite}}{\text{adjacent}} = \frac{b}{a}$$

Since the hypotenuse of the right triangle is always 1 unit long, the values of the x and y coordinates of a point on the circle are always equal to the cosine and sine (respectively) of the angle θ.

This angle is measured in a unit called radians, which corresponds to the distance around the unit circle from the point (1,0). The circumference of the unit circle is 2π, so 2π radian is the same as 360°. Any other angle less than 360° can be represented as some fraction of 2π radians. For example, A 90° angle is the same as $\frac{1}{4}$ of the way around the circle, which would be $\frac{2\pi}{4} = \frac{\pi}{2}$.

Some possible values of θ are listed below, along with their corresponding values of sine and cosine.

angle measure, θ	sin θ	cos θ
0	0	1
$\frac{\pi}{6}$	$\frac{1}{2}$	$\frac{\sqrt{3}}{2}$
$\frac{\pi}{4}$	$\frac{\sqrt{2}}{2}$	$\frac{\sqrt{2}}{2}$
$\frac{\pi}{3}$	$\frac{\sqrt{3}}{2}$	$\frac{1}{2}$
$\frac{\pi}{2}$	1	0

The trigonometry we are familiar with so far is based on only right triangles and acute angles. However, with help of Unit Circle we can extend our understanding of trigonometric functions plus also become familiar with the use of non-acute angles.

Angles in the Unit Circle

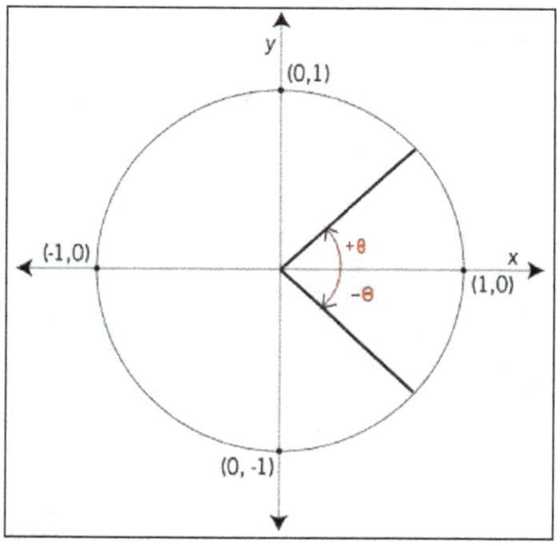

An angle on Unit Circle is always measured from the positive xx-axis, with its vertex at the origin. It is measured to a point on the unit circle. The ray that begins at the origin and contains the point on the unit circle is called the terminal side.

An angle is said to be positive if it is measured by going in anticlockwise direction from the positive x-axis and negative if it is measured by going in clockwise direction from the x-axis.

Since $2\pi \text{rad} = 360°$, any degree measurement can be converted to radians, and vice versa.

Let d be an angle's measurement in degrees, and let r be that same angle's measurement in radians.

$$r = \frac{\pi d}{180}$$
$$d = \frac{180r}{\pi}.$$

Coordinates in the Unit Circle

A right triangle AOBAOB with right angle at AA lies on the Cartesian plane such that OA lies on the x-axis, point O lies on the origin and point BB lies anywhere on the Unit Circle. Note that OB=1 units.

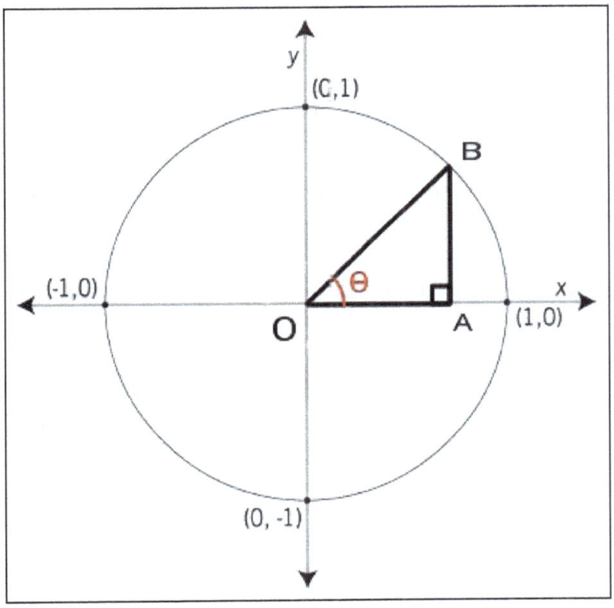

The sine and cosine trigonometric functions are given below. When defining these functions in terms of the unit circle, it is possible to have negative lengths. If OA is along the negative x-axis, then OA is considered to be negative. Likewise, if AB extends below the x-axis, then AB is considered to be negative.

$$\sin(\theta) = \frac{\text{opposite}}{\text{hypotenuse}} = \frac{AB}{OB} = \frac{AB}{1} = AB$$
$$\cos(\theta) = \frac{\text{adjacent}}{\text{hypotenuse}} = \frac{OA}{OB} = \frac{OA}{1} = OA.$$

By this convention, the sine of an angle is considered to be the y-coordinate of a point on the unit circle given by that angle. Likeways, the cosine of an angle is considered to be the x-coordinate of a point on the unit circle given by that angle. In general, to compute the sine or cosine of any angle θ, look at the coordinates of the point on unit circle made by that angle.

Special Angles on the Unit Circle

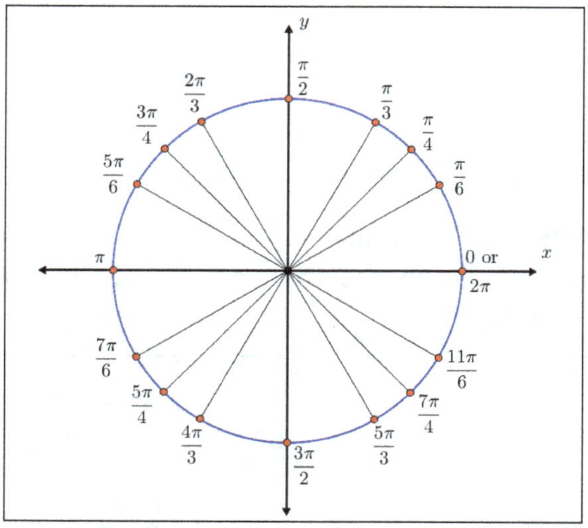

The sixteen special angles (measured in radians) on the unit circle, each labeled at the terminal point.

The special angles are angles on the unit circle for which the coordinates are well-known. These coordinates can be solved for with right-triangle relationships.

2

Trigonometric Functions

Trigonometric functions are the angle functions which gives a relationship between an angle of a right-angles triangle and ratios of two side lengths. It includes sine, cosine, tangent, cosecant, secant and cotangent. This is an introductory chapter which will briefly introduce about these trigonometric functions.

Sine

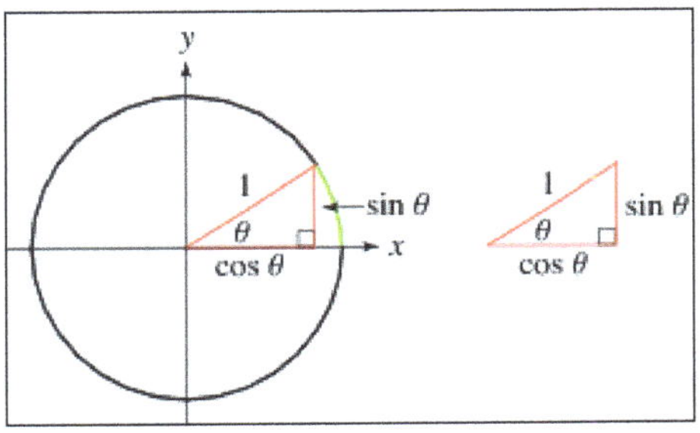

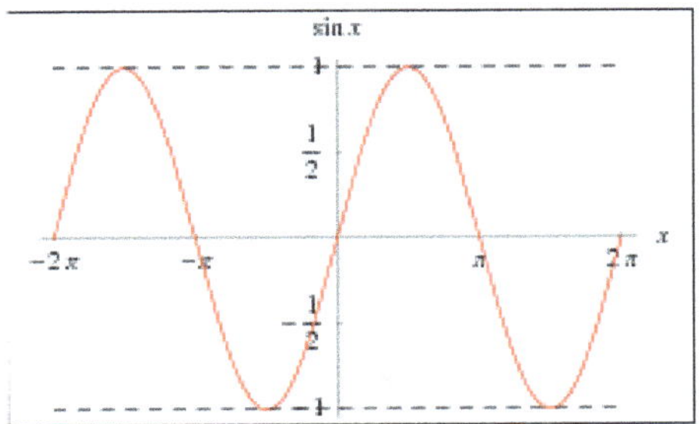

The sine function sin x is one of the basic functions encountered in trigonometry (the others being the cosecant, cosine, cotangent, secant, and tangent). Let theta be an angle measured counterclockwise from the x-axis along an arc of the unit circle. Then sin θ is the vertical coordinate of the arc endpoint, as illustrated in the left figure above.

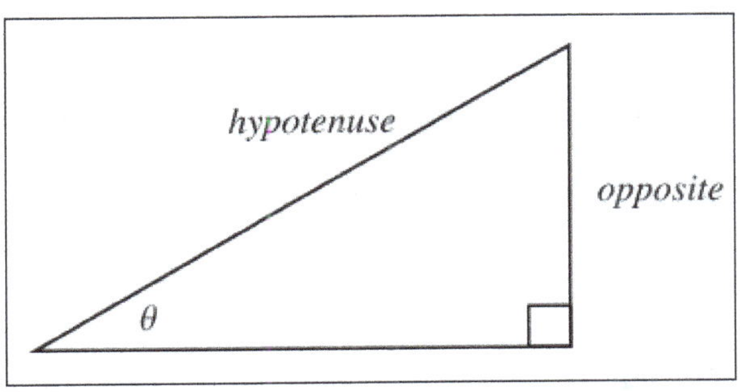

The common schoolbook definition of the sine of an angle θ in a right triangle (which is equivalent to the definition just given) is as the ratio of the lengths of the side of the triangle opposite the angle and the hypotenuse, i.e.,

$$\sin\theta = \frac{\text{opposite}}{\text{hypotenuse}}.$$

A convenient mnemonic for remembering the definition of the sine, as well as the cosine and tangent, is SOHCAHTOA (sine equals opposite over hypotenuse, cosine equals adjacent over hypotenuse, tangent equals opposite over adjacent).

As a result of its definition, the sine function is periodic with period 2π. By the Pythagorean theorem, sin θ also obeys the identity:

$$\mathrm{Sin}^2\,\theta + \mathrm{Cos}^2\,\theta = 1$$

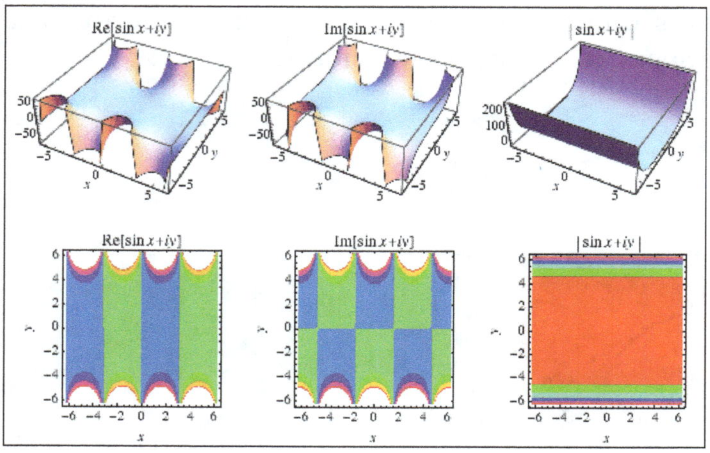

The definition of the sine function can be extended to complex arguments z, illustrated above, using the definition,

$$\sin z = \frac{e^{iz} - e^{-iz}}{2i}$$
$$= \frac{1}{2} i \left(e^{-iz} - e^{iz}\right),$$

where e is the base of the natural logarithm and i is the imaginary number. Sine is an entire function and is implemented in the Wolfram Language as Sin[z].

A related function known as the hyperbolic sine is similarly defined,

$$\sin z = \frac{1}{2}\left(e^{z} - e^{-z}\right).$$

The sine function can be defined analytically by the infinite sum,

$$\sin x = \sum_{n=1}^{\infty} \frac{(-1)^{n-1}}{(2n-1)!} x^{2n-1}.$$

It is also given by the imaginary part of the complex exponential,

$$\sin x = \mathrm{I}\left[e^{ix}\right].$$

The multiplicative inverse of the sine function is the cosecant, defined as:

$$\csc x = \frac{1}{\sin x}.$$

The sine function is also given by the limit,

$$\sin(z) = -\pi \lim_{n\to\infty} \frac{1}{\operatorname{In} n} \sum_{k=1}^{\infty} \frac{\mu(k)}{k} \operatorname{In}\left(\frac{n}{k}\right) \operatorname{frac}\left(\frac{kz}{2\pi}\right),$$

where μ (k) is the Möbius function and frac (x) is the fractional part.

The derivative of sin x is:

$$\frac{d}{dx} \sin x = \cos x,$$

and its indefinite integral is:

$$\int \sin x\, dx = -cos\, x + C,$$

where C is a constant of integration.

Using the results from the exponential sum formulas:

$$\begin{aligned}
\sum_{n=0}^{N} \sin(nx) &= \operatorname{I}\left[\sum_{n=0}^{N} e^{inx}\right] \\
&= \operatorname{I}\left[\frac{e^{i(N+1)x} - 1}{e^{ix} - 1}\right] \\
&= \operatorname{I}\left[\frac{e^{i(N+1)x/2}}{e^{ix/2}} \frac{e^{i(N+1)x/2} - e^{i(N-1)x/2}}{e^{ix/2} - e^{ix/2}}\right] \\
&= \frac{\sin\left(\frac{1}{2}(N+1)x\right)}{\sin\left(\frac{1}{2}x\right)} \operatorname{I}\left[e^{iNx/2}\right] \\
&= \frac{\sin\left(\frac{1}{2}Nx\right)\sin\left[\frac{1}{2}(N+1)x\right]}{\sin\left(\frac{1}{2}x\right)}.
\end{aligned}$$

Similarly,

$$\sum_{n=0}^{\infty} \mathrm{p}^n \sin(nx) = \operatorname{I}\left[\sum_{n=0}^{\infty} \mathrm{p}^n e^{inx}\right]$$

$$= \mathrm{I}\left[\frac{1-pe^{-ix}}{1-2p\cos x+p^2}\right]$$

$$= \frac{p\sin x}{1-2p\cos x+p^2}.$$

The sum of $\sin^2$ (k x) can also be done in closed form,

$$\sum_{k=0}^{N} \sin^2(kx) = \frac{1}{4}\{1+2N-\csc x\sin[x(1+2N)]\}.$$

A related sum identity is given by,

$$\sum_{k=1}^{n=1} \sin\left(\frac{k\pi}{n}\right) = \cot\left(\frac{\pi}{2n}\right)$$

Product identities include,

$$\pi s\prod_{n=1}^{\infty}\left(1-\frac{s^2}{n^2}\right) = \sin(\pi s),$$

which is more commonly written as an identity for the sinc function or in the form,

$$\sin x = x\prod_{n=1}^{\infty}\left(1-\frac{x^2}{n^2\pi^2}\right)$$

Another product formula is,

$$\prod_{k=1}^{n-1}\sin\left(\frac{k\pi}{n}\right) = 2^{1-n}\,n,$$

The sine function obeys the identity:

$$\sin(n\,\theta) = 2\cos\theta\sin[(n-1)\theta] - \sin[(n-2)\theta]$$

and the multiple-angle formula:

$$\sin(n\,x) = \sum_{k=0}^{n}\binom{n}{k}\cos^k x\sin^{n-k} x\sin\left[\frac{1}{2}(n-k)\pi\right],$$

Where $\binom{n}{k}$ is a binomial coefficient. It is related to tan(x/2) via:

$$\sin x = \frac{2\tan\left(\frac{1}{2}x\right)}{1+\tan^2\left(\frac{1}{2}x\right)}.$$

A curious identity is given by,

$$\frac{\sin(n\alpha)}{\sin\alpha}=\sum_{j=1}^{n}\prod_{\substack{k=1\\k\neq j}}^{n}\frac{\sin\left(\alpha+\theta_j-\theta_k\right)}{\sin\left(\theta_j-\theta_k\right)}$$

for all α and $\theta_j \neq \theta_k$.

Cvijović and Klinowski show that the sum:

$$S_V(\alpha)=\sum_{k=0}^{\infty}\frac{\sin(2k+1)\alpha}{(2k+1)^{\nu}}$$

has closed form for v=2 n+1,

$$S_{2n+1}(\alpha)=\frac{(-1)^n}{4(2n)!}\pi^{2n+1}E_{2n}\left(\frac{\alpha}{\pi}\right),$$

where E_n (x) is an Euler polynomial.

A continued fraction representation of sin x is:

$$\sin x \frac{x}{1+\cfrac{x^2}{\left(2\cdot 3-x^2\right)+\cfrac{2\cdot 3x^2}{\left(4\cdot 5-x^2\right)+\cfrac{4\cdot 5x^2}{\left(6\cdot 7-x^2\right)+\dots}}}}$$

The value of sin(2 π/n) is irrational for all integers n>1 except 2, 4, and 12, for which sin (π)=0, sin(π/2)=1, and sin(π/6)=1/2, respectively, a result that is essentially known as Niven's theorem.

The Fourier transform of sin($2\pi k_0$ x) is given by,

$$\begin{aligned}F\left[\sin(2\pi k_0 x)\right](k) &\int_{-\infty}^{\infty} e^{\quad ikx}\sin(2\pi k_0 x)\,dx\\ &=-i\left[\delta(k+k_0)-\delta(k-k_0)\right]\end{aligned}$$

A definite integral involving sin x is given by,

$$\int_0^{\infty}\sin\left(x^n\right)dx=\Gamma\left(1+\frac{1}{n}\right)\sin\left(\frac{\pi}{2n}\right)$$

for n>1 where Γ (z) is the gamma function.

Sine Function is Odd

Theorem:

For all $z \in \mathbb{C}$:

$$\sin(-z) = -\sin z$$

That is, the sine function is odd.

Proof:

From the definition of the sine function:

$$\sin z = \sum_{n=0}^{\infty} (-1)^n \frac{z^{2n+1}}{(2n+1)!}$$
$$= z - \frac{z^3}{3!} + \frac{z^5}{5!} - \cdots$$

From Sign of Odd Power, we have that:

$$\forall n \in \mathbb{N} : -(z^{2n+1}) = (-z)^{2n+1}$$

The result follows directly.

Derivative of Sine Function

Theorem:

$$D_x (\sin x) = \cos x$$

Corollary:

$$D_x (\sin ax) = a \cos ax$$

Proof:

From the definition of the sine function, we have:

$$\sin x = \sum_{n=0}^{\infty} (-1)^n \frac{x^{2n+1}}{(2n+1)!}$$

From Radius of Convergence of Power Series over Factorial, this series converges for all x.

From Power Series is Differentiable on Interval of Convergence:

$$D_x(\sin x) = \sum_{n=0}^{\infty} (-1)^n (2n+1) \frac{x^{2n}}{(2n+1)!}$$
$$= \sum_{n=0}^{\infty} (-1)^n \frac{x^{2n}}{(2n)!}$$

The result follows from the definition of the cosine function.

Proof:

$$D_x(\sin x) = \lim_{h\to 0} \frac{\sin(x+h) - \sin(x)}{h}$$ Definition of Derivative of Real Function at Point

$$= \lim_{h\to 0} \frac{\sin(x)\cos(h) + \sin(h)\cos(x) - \sin(x)}{h}$$ Sine of Sum

$$= \lim_{h\to 0} \frac{\sin(x)\cos(h-1) + \sin(h)\cos(x)}{h}$$ collecting terms containing $\sin(x)$ and factoring

$$= \lim_{h\to 0} \frac{\sin(x)\cos(h) - 1}{h} + \lim_{h\to 0} \frac{\sin(h)\cos(x)}{h}$$ Sum Rule for Limits of Functions

$$= \sin(x) \times 0 + 1 \times \cos(x)$$ Limit of Sine of X over X and Limit of $(\text{Cosine}(X) - 1)$ over X

$$= \cos(x)$$

Proof:

$$D_x \sin x = D_x \cos\left(\frac{\pi}{2} - x\right)$$ Cosine of Complement equals Sine

$$= \sin\left(\frac{\pi}{2} - x\right)$$ Derivative of Cosine Function and Chain Rule

$$= \cos x$$ Sine of Complement equals Cosine

Proof:

$$D_x \sin x = \lim_{h\to 0} \frac{\sin(x+h) - \sin(x)}{h}$$ Definition of Derivative of Real Function at Point

$$= \lim_{h\to 0} \frac{\sin\left(\left(x+\frac{h}{2}\right)+\frac{h}{2}\right)-\sin\left(\left(x+\frac{h}{2}\right)-\frac{h}{2}\right)}{h}$$

$$= \lim_{h\to 0} \frac{2cos\left(x+\frac{h}{2}\right)\sin\left(\frac{h}{2}\right)}{h}$$ Simpson's Formula for Cosine by Sine

$$= \lim_{h\to 0} \cos\left(x+\frac{h}{2}\right)\lim_{h\to 0}\frac{\sin\left(\frac{h}{2}\right)}{\frac{h}{2}}$$ Multiple Rule for Limits of Functions and Product Rule for Limits of Functions

$$= \cos x \times 1$$ Continuity of Cosine and Limit of Sine of X over X

$$= \cos x$$

Proof:

$$\arcsin(x) = \int_0^x \frac{dx}{\sqrt{1-x^2}}$$ Arcsin as an Integral

$$\leadsto \quad \frac{d(\arcsin(y))}{dy} = \frac{d\left(\int_0^y \frac{1}{\sqrt{1-y^2}}dy\right)}{dy}$$

$$= \frac{1}{\sqrt{1-y^2}}$$ Corollary to Fundamental Theorem of Calculus : First Part

We get the same answer as Derivative of Arcsine Function.

By definition of real arcsin function, arcsin is bijective on its domain [−1 . . 1].

Thus its inverse is itself a mapping.

From Inverse of Inverse of Bijection, its inverse is the sin function.

$$\frac{\mathrm{d}(\sin\theta)}{\mathrm{d}\theta} = 1/\frac{1}{\sqrt{1-\sin^2\theta}}$$ Derivative of Inverse Function

$$= \pm\sqrt{1-\sin^2\theta}$$ Positive in Quadrant I and Quadrant IV, Negative Quadrant II and Quadrant III

$$\frac{\mathrm{d}(\sin\theta)}{\mathrm{d}\theta} = \cos\theta$$

Power Series Expansion for Sine Function

Theorem:

The sine function has the power series expansion:

$$\sin x = \sum_{n=0}^{\infty} (-1)^n \frac{x^{2n+1}}{(2n+1)!}$$
$$= x - \frac{x^3}{3!} + \frac{x^5}{5!} - \frac{x^7}{7!} \cdots$$

valid for all $x \in \mathbb{R}$.

Proof:

From Derivative of Sine Function:

$$\frac{d}{dx}\sin x = \cos x$$

From Derivative of Cosine Function

$$\frac{d}{dx}\cos x = -\sin x$$

Hence:

$$\frac{d^2}{dx^2}\sin x = -\sin x$$
$$\frac{d^3}{dx^3}\sin x = -\cos x$$
$$\frac{d^4}{dx^4}\sin x = \sin x$$

and so for all $m \in \mathbb{N}$:

$$m = 4k: \ \frac{d^m}{dx^m}\sin x = \sin x$$

$$m = 4k+1: \quad \frac{d^m}{dx^m}\sin x = \cos x$$

$$m = 4k+2: \quad \frac{d^m}{dx^m}\sin x = -\sin x$$

$$m = 4k+3: \quad \frac{d^m}{dx^m}\sin x = -\cos x$$

Where $k \in \mathbb{Z}$.

This leads to the Maclaurin series expansion:

$$\sin x = \sum_{r=0}^{\infty}\left(\frac{x^{4k}}{(4k)!}\sin(0) + \frac{x^{4k+1}}{(4k+1)!}\cos(0) - \frac{x^{4k+2}}{(4k+2)!}\sin(0) - \frac{x^{4k+3}}{(4k+3)!}\cos(0)\right)$$

$$= \sum_{r=0}^{\infty}\left(\frac{x^{4k+1}}{(4k+1)!} - \frac{x^{4k+3}}{(4k+3)!}\right) \qquad \text{Sine of Zero is Zero, Cosine of Zero is One}$$

$$= \sum_{n=0}^{\infty}(-1)^n \frac{x^{2k+1}}{(2n+1)!} \qquad \text{setting } n = 2k$$

From Series of Power over Factorial Converges, it follows that this series is convergent for all x.

Complex Sine Function

The complex function $\sin : \mathbb{C} \to \mathbb{C}$ is defined as:

$$\sin z = \sum_{n=0}^{\infty}(-1)^n \frac{z^{2n+1}}{(2n+1)!}$$

$$= z - \frac{z^3}{3!} + \frac{z^5}{5!} - \frac{z^7}{7!} + \cdots + (-1)^n \frac{z^{2n+1}}{(2n+1)!} + \cdots$$

Complex Sine Function is Entire

Theorem

Let $\sin : \mathbb{C} \to \mathbb{C}$ be the complex sine function.

Then sin is entire.

Proof:

By the definition of the complex sine function, sin admits a power series expansion about 0:

$$\sin z = \sum_{n=0}^{\infty}(-1)^n \frac{z^{2n+1}}{(2n+1)!}$$

By Complex Function is Entire iff it has Everywhere Convergent Power Series, to show that sin is entire it suffices to show that this series is everywhere convergent.

From Radius of Convergence from Limit of Sequence: Complex Case, it is sufficient to show that:

$$\lim_{n\to\infty}\left|\frac{(-1)^{n+1}}{(2n+3)!}\times\frac{(2n+1)!}{(-1)n}\right|=0$$

We have:

$$\begin{aligned}\lim_{n\to\infty}\left|\frac{(-1)^{n+1}}{(2n+3)!}\times\frac{(2n+1)!}{(-1)^n}\right| &= |-1|\lim_{n\to\infty}\left|\frac{(2n+1)!}{(2n+3)(2n+2)(2n+1)!}\right| \\ &= \lim_{n\to\infty}\left|\frac{1}{(2n+3)(2n+2)}\right| \\ &= 0\end{aligned}$$

hence the result.

Proof:

Let:

$$\begin{aligned} f(z) &= \exp z \\ g(z) &= iz \\ h(z) &= -iz \end{aligned}$$

for all $z \in \mathbb{C}$.

By Complex Exponential Function is Entire, we have that f is entire.

By Polynomial is Entire, we have that g and h are entire.

Therefore, by Composition of Entire Functions is Entire, we have that both f ∘ g and f ∘ h are entire.

By Linear Combination of Entire Functions is Entire, we then have that:

$$\frac{1}{2i}(f\circ g-f\circ h)$$

is entire.

Note that:

$$\frac{1}{2i}\left(\left(f\circ g\right)\left(z\right)-\left(f\circ h\right)\left(z\right)\right)=\frac{1}{2i}\left(\exp\left(iz\right)-\exp\left(iz\right)\right) \qquad \text{Sine Exponential Formulation}$$
$$=\sin z$$

Therefore, sin is an entire function.

Complex Sine Function is Unbounded

Theorem:

Let $\sin:\mathbb{C}\to\mathbb{C}$ be the complex sine function.

Then sin is unbounded.

Proof:

By Complex Sine Function is Entire, we have that sin is a entire function.

Aiming for a contradiction, suppose that sin was a bounded function.

Then, by Liouville's Theorem, we would have that sin is a constant function.

However we have, for instance, by Sine of Zero is Zero:

$$\sin 0=0$$

and by Sine of 90 Degrees:

$$\sin\frac{\pi}{2}=1$$

Therefore, sin is clearly not a constant function, a contradiction.

We hence conclude, by Proof by Contradiction, that sin is unbounded.

Sine Exponential Formulation

Theorem:

For any complex number z:

$$\sin z=\frac{\exp(iz)-\exp(-iz)}{2i}$$

$\exp z$ denotes the exponential function

$\sin z$ denotes the complex sine function

i denotes the inaginary unit.

Real Domain

This result is often presented and proved separately for arguments in the real domain:

$$\sin x = \frac{e^{ix} - e^{-ix}}{2i}$$

Proof:

From the definition of the sine function:

$$\begin{aligned}\sin z &= \sum_{n=0}^{\infty} (-1)^n \frac{z^{2n+1}}{(2n+1)!} \\ &= z - \frac{z^3}{3!} + \frac{z^5}{5!} - \frac{z^7}{7!} + \cdots + (-1)^n \frac{z^{2n+1}}{(2n+1)!} + \cdots\end{aligned}$$

From the definition of the exponential as a power series:

$$\begin{aligned}\exp z &= \sum_{n=0}^{\infty} \frac{z^n}{n!} \\ &= 1 + \frac{z}{1!} + \frac{z^2}{2!} - \frac{z^3}{3!} + \cdots \frac{z^n}{n!} + \cdots\end{aligned}$$

Then, starting from the right hand side:

$$\begin{aligned}\frac{\exp(iz) - \exp(-ix)}{2i} &= \frac{1}{2i}\left(\sum_{n=0}^{\infty} \frac{(iz)^n}{n!} - \sum_{n=0}^{\infty} \frac{(-iz)^n}{n!}\right) \\ &= \frac{1}{2i}\sum_{n=0}^{\infty}\left(\frac{(iz)^n - (-iz)^n}{n!}\right) \\ &= \frac{1}{2i}\sum_{n=0}^{\infty}\left(\frac{(iz)^{2n} - (-iz)^{2n}}{(2n!)} + \frac{(iz)^{2n+1} - (-iz)^{2n+1}}{(2n+1)!}\right) && \text{split into even and odd } n \\ &= \frac{1}{2i}\sum_{n=0}^{\infty}\left(\frac{(iz)^{2n+1} - (-iz)^{2n+1}}{(2n+1)!}\right) && as\,(-iz)^{2n} = (iz)^{2n} \\ &= \frac{1}{2i}\sum_{n=0}^{\infty}\frac{2(iz)^{2n+1}}{(2n+1)!} && as\,(-1)^{2n+1} = -1 \\ &= \frac{1}{i}\sum_{n=0}^{\infty}\frac{(iz)^{2n+1}}{(2n+1)!} && \text{cancel } 2 \\ &= \frac{1}{i}\sum_{n=0}^{\infty}\frac{i(-1)^n z^{2n+1}}{(2n+1)!} && as\, i^{2n+1} = i(-1))^n\end{aligned}$$

$$= \sum_{n=0}^{\infty} (-1)^n \frac{z^{2n+1}}{(2n+1)!} \qquad \text{cancel } i$$
$$= \sin z$$

Proof:

From Euler's Formula:

$$\exp(iz) = \cos z + i \sin z$$

Then, starting from the right hand side:

$$\begin{aligned}
\frac{\exp(iz) - \exp(iz)}{2i} &= \frac{(\cos z + i\sin z) - (\cos(-z) + i\sin(-z))}{2i} \\
&= \frac{(\cos z + i\sin z - \cos z - i\sin(-z))}{2i} && \text{Cosine Function is Even} \\
&= \frac{i\sin z - i\sin(-z)}{2i} \\
&= \frac{i\sin z - i(-\sin(-z))}{2i} && \text{Sine Function is Odd} \\
&= \frac{2i\sin z}{2i} \\
&= \sin z
\end{aligned}$$

Proof:

$$\begin{aligned}
\exp(iz) &= \cos z + i\sin z && \text{Euler's Formula} \\
\exp(-iz) &= \cos z - i\sin z && \text{Euler's Formula : Corollary} \\
\leadsto \exp(iz) - \exp(-iz) &= (\cos z + i\sin z) - (-\cos z - i\sin z) \\
&= 2i\sin z && \text{simplifying} \\
\leadsto \frac{\exp(iz) - \exp(-iz)}{2i} &= \sin z
\end{aligned}$$

Also presented as:

This result can also be presented as:

$$\sin \mathrm{z} = \frac{1}{2} i \left(e^{-iz} - e^{iz}\right)$$

Laplace Transform of Sine

Theorem:

Let sin denote the real sine function.

Let L{f} denote the Laplace transform of a real function f.

Then:

$$L\{\sin at\} = \frac{a}{s^2 + a^2}$$

Where $a \in \mathbb{R}_{>0}$ is constant, and $\Re(s) > a$.

Proof:

$$L\{\sin at\}(s) = \int_0^{\to +\infty} e^{-st} \sin at\, dt$$ Definition of Laplace Transform

$$= \lim_{L \to \infty} \int_0^L e^{-st} \sin at\, dt$$ Definition of Improper Integral

$$= \lim_{L \to \infty} \left[\frac{e^{-st}(-s \sin at - a \cos at)}{(-s)^2 + a^2} \right]_0^L$$ primitive of $e^{ax} \sin bx$

$$= \lim_{L \to \infty} \left(\frac{e^{-st}(-s \sin aL - a \cos aL)}{s^2 + a^2} - \frac{e^{-s \times 0}(-s \sin(0 \times a) - a \cos(0 \times a))}{s^2 + a^2} \right)$$

$$= \lim_{L \to \infty} \left(\frac{s \sin 0 + a \cos 0}{s^2 + a^2} - \frac{e^{-sL}(s \sin aL + a \cos aL)}{s^2 - a^2} \right)$$ Exponential of Zero and rearranging

$$= \frac{s \sin 0 + a \cos 0}{s^2 + a^2} - 0$$ Exponential Tends to Zero

$$= \frac{a}{s^2 + a^2}$$ Sine of Zero is Zero, Cosine of Zero is One

Proof:

$$\mathcal{L}\{e^{iat}\} = \frac{1}{s - ia}$$ Laplace Transform of Exponential

$$= \frac{s + ia}{s^2 + a^2}$$ multiplying top and bottom by $s + ia$

Also:

$$\mathcal{L}\{e^{iat}\} = \mathcal{L}\{\cos at + i \sin at\}$$ Euler's Formula

$$= \mathcal{L}\{\cos at\} + i\,\mathcal{L}\{\sin at\}$$ Linear Combination of Laplace Transforms

So:

$$\mathcal{L}\{\sin at\} = \mathcal{J}m\left(L\left\{e^{iat}\right\}\right)$$
$$= \mathcal{J}m\left(\frac{s+ia}{s^2+a^2}\right)$$
$$= \frac{a}{s^2+a^2}$$

Proof:

$$\mathcal{L}\{\sin at\} = \mathcal{L}\left\{\frac{e^{iat} - e^{-iat}}{2i}\right\} \quad \text{Sine Exponential Formulation}$$
$$= \frac{1}{2i}\left(\mathcal{L}\left\{e^{at}\right\} - \mathcal{L}\left\{e^{-at}\right\}\right) \quad \text{Linear Combination of Laplace Transforms}$$
$$= \frac{1}{2i}\left(\frac{1}{s-ia} - \frac{1}{s+ia}\right) \quad \text{Laplace Transform of Exponential}$$
$$= \frac{1}{2i}\left(\frac{s+ia-s+ia}{s^2+a^2}\right) \quad \text{simplifying}$$
$$= \frac{1}{2i}\left(\frac{2ia}{s^2+a^2}\right) \quad \text{simplifying}$$
$$= \frac{a}{s^2+a^2} \quad \text{simplifying}$$

Proof:

By definition of the Laplace transform:

$$L\{\sin at\} = \int_0^{\to +\infty} e^{-st} \sin \mathrm{at}\, \mathrm{dt}$$

From Integration by Parts:

$$\int fg'dt = fg - \int f'gdt$$

Here:

$$f = \sin at$$
$$\rightsquigarrow \quad f' = a\cos at \quad \text{Derivative of } \sin ax$$
$$g' = e^{-st}$$
$$\rightsquigarrow \quad g = -\frac{1}{s}e^{-st} \quad \text{Primitive of } e^{ax}$$

So:

$$\int e^{-st}\sin at\,dt = -\frac{1}{s}e^{-st}\sin at + \frac{a}{s}\int e^{-st}\cos at\,dt$$

Consider:

$$\int e^{-st}\cos at\,dt$$

Again, using Integration by Parts:

$$\int hj'\,dt = hj\int h'j\,dt$$

Here:

$$h = \cos at$$

$$\rightsquigarrow\ h' = -a\sin at \quad \text{Derivative of Cosine Function}$$

$$j' = e^{-st}$$

$$\rightsquigarrow\ j = -\frac{1}{s}e^{-st} \quad \text{Primitive of Exponential Function}$$

So:

$$\int e^{-st}\cos at\,dt = -\frac{1}{s}e^{-st}\cos at - \frac{a}{s}\int e^{-st}\sin at\,dt$$

Substituting this into: $\int e^{-st}\sin at\,dt = -\frac{1}{s}e^{-st}\sin at + \frac{a}{s}\int e^{-st}\cos at\,dt$:

$$\begin{aligned}\int e^{-st}\sin at\,dt &= -\frac{1}{s}e^{-st}\sin at + \frac{a}{s}\left(-\frac{1}{s}e^{-st}\cos at - \frac{a}{s}\int e^{-st}\sin at\,dt\right)\\ &= -\frac{1}{s}e^{-st}\sin at + \frac{a}{s^2}e^{-st}\cos at - \frac{a^2}{s^2}\int e^{-st}\sin at\,dt\end{aligned}$$

$$\rightsquigarrow\ \left(1+\frac{a^2}{s^2}\right)\int e^{-st}\sin at\,\mathrm{d}t = -e^{-st}\left(\frac{1}{s}\sin at + \frac{a}{s^2}\cos at\right)$$

Evaluating at $t = 0$ and $t \to +\infty$:

$$\begin{aligned}\left(1+\frac{a^2}{s^2}\right)L\{\sin at\} &= \left[-e^{-st}\left(\frac{1}{s}\sin at + \frac{a}{s^2}\cos at\right)\right]_{t=0}^{t\to+\infty}\\ &= 0 - \left(-1\left(\frac{1}{s}\times 0 + \frac{a}{s^2}\times 1\right)\right) \quad \text{Boundedness of Real Sine and Cosine, Complex Exponential Tends to Zero}\\ &= \frac{a}{s^2}\end{aligned}$$

$$\rightsquigarrow \qquad L\{\sin at\} = \frac{a}{s^2}\left(1+\frac{a^2}{s^2}\right)^{-1}$$
$$= \frac{a}{s^2}\left(\frac{s^2}{a^2+s^2}\right)$$
$$= \frac{a}{s^2+a^2}$$

Proof:

From Laplace Transform of Second Derivative:

$$\mathcal{L}\{f''(t)\} = s^2\mathcal{L}\{f(t)\} - s\,f(0) - f'(0)$$

under suitable conditions.

Then:

$$f(t) = \sin at$$
$$\rightsquigarrow \quad f'(t) = a\cos at$$
$$f''(t) = -a^2 \sin at$$
$$f(0) = 0$$
$$f'(0) = a$$

from equation $\mathcal{L}\{f''(t)\} = s^2\mathcal{L}\{f(t)\} - s\,f(0) - f'(0)$ substituting for $f(t)$, $f'(0)$ and $f(0$

$$\rightsquigarrow \quad \mathcal{L}\{-a^2\sin at\} = s^2\mathcal{L}\{\sin at\} - s\times 0 - a \qquad \text{rearranging}$$

$$\rightsquigarrow \quad -a^2\mathcal{L}\{\sin at\} = s^2\mathcal{L}\{\sin at\} - a$$

$$\rightsquigarrow \quad \mathcal{L}\{\sin at\} = \frac{a}{s^2+a^2}$$

Cosine

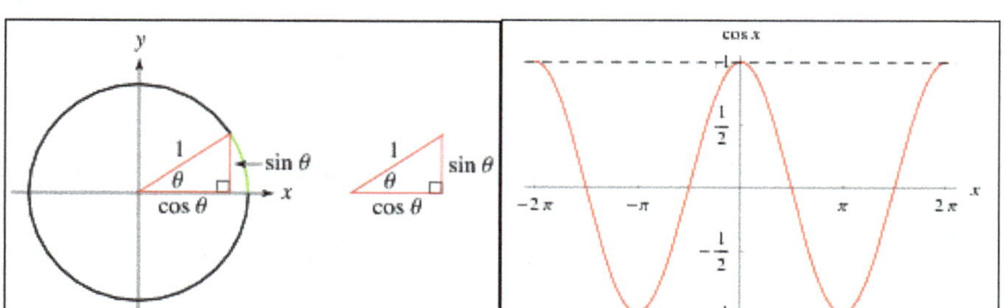

The cosine function cos x is one of the basic functions encountered in trigonometry (the others being the cosecant, cotangent, secant, sine, and tangent). Let theta be an angle measured counterclockwise from the x-axis along the arc of the unit circle. Then cos θ is the horizontal coordinate of the arc endpoint.

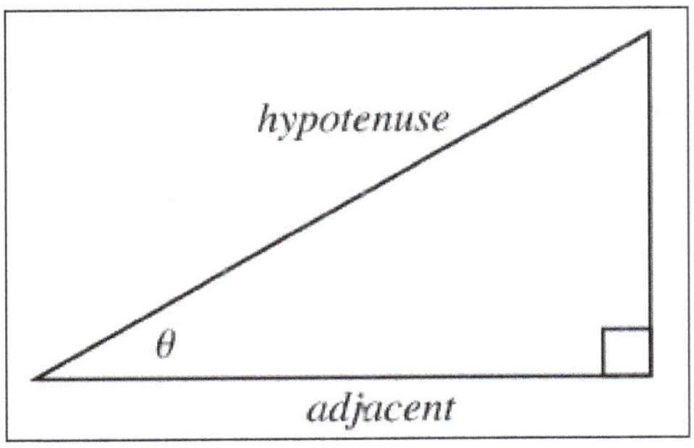

The common schoolbook definition of the cosine of an angle θ in a right triangle (which is equivalent to the definition just given) is as the ratio of the lengths of the side of the triangle adjacent to the angle and the hypotenuse, i.e.

$$\cos\theta = \frac{\text{adjacent}}{\text{hypotenuse}}.$$

A convenient mnemonic for remembering the definition of the sine, cosine, and tangent is SOHCAHTOA (sine equals opposite over hypotenuse, cosine equals adjacent over hypotenuse, tangent equals opposite over adjacent).

As a result of its definition, the cosine function is periodic with period 2π. By the Pythagorean theorem, cos θ also obeys the identity,

$$\sin^2\theta + \cos^2\theta = 1.$$

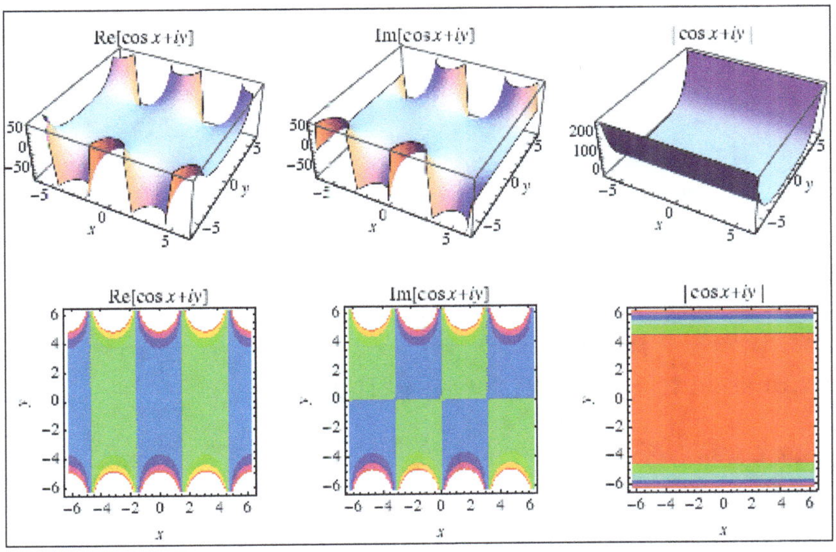

The definition of the cosine function can be extended to complex arguments z using the definition,

$$\cos z = \frac{1}{2}\left(e^{iz} + e^{iz}\right),$$

where e is the base of the natural logarithm and i is the imaginary number. Cosine is an entire function and is implemented in the Wolfram Language as Cos[z].

A related function known as the hyperbolic cosine is similarly defined,

$$\cos z = \frac{1}{2}\left(e^{z} + e^{-z}\right).$$

The cosine function has a fixed point at 0.739085..., a value sometimes known as the Dottie number.

The cosine function can be defined analytically using the infinite sum:

$$\begin{aligned}\cos x &\equiv \sum_{n=0}^{\infty} \frac{(-1)^n x^{2n}}{(2n)!} \\ &= 1 - \frac{x^2}{2!} + \frac{x^4}{4!} - \frac{x^6}{6!} + \ldots,\end{aligned}$$

or the infinite product:

$$\cos x = \prod_{n=1}^{\infty}\left[1 - \frac{4x^2}{\pi^2(2n-1)^2}\right]$$

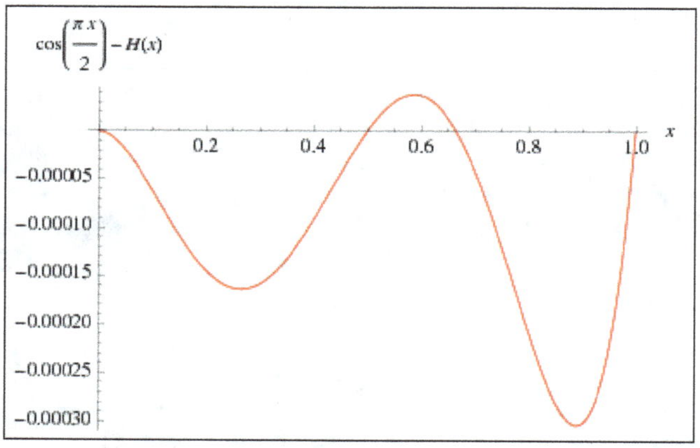

A close approximation to $\cos(\pi x/2)$ for $x \in [0,1]$ is,

$$H(x) = 1 - \frac{x^2}{x + (1-x)\sqrt{\frac{2-x}{3}}}$$

$$\approx \cos\left(\frac{\pi}{2}x\right)$$

where the difference between $cos(\pi/2)$ and Hardy's approximation is plotted above.

The cosine obeys the identity:

$$\cos(nx) = \sum_{k=0}^{n} \binom{n}{k} \cos^k x \sin^{n-k} x \cos\left[\frac{1}{2}(n-k)\pi\right].$$

where $\binom{n}{k}$ is a binomial coefficient. It is related to $\tan(x/2)$ via,

$$\cos x = \frac{1-\tan^2\left(\frac{1}{2}x\right)}{1+\tan^2\left(\frac{1}{2}x\right)}$$

Summation of $\cos(nx)$ from n=0 to N can be done in closed form as,

$$\begin{aligned}
\sum_{n=0}^{N} \cos(nx) &= \Re\left[\sum_{n=0}^{N} e^{inx}\right] \\
&= \Re\left[\frac{e^{i(N+1)x}-1}{e^{ix}-1}\right] \\
&= \Re\left[\frac{e^{i(N+1)x/2}}{e^{ix/2}} \frac{e^{i(N+1)x/2}-e^{-i(N+1)x/2}}{e^{ix/2}-e^{ix/2}}\right] \\
&= \frac{\sin\left[\frac{1}{2}(N+1)x\right]}{\sin\left(\frac{1}{2}x\right)} \Re\left[e^{iNx/2}\right] \\
&= \frac{\cos\left(\frac{1}{2}Nx\right)\sin\left[\frac{1}{2}(N+1)x\right]}{\sin\left(\frac{1}{2}x\right)}.
\end{aligned}$$

Similarly,

$$\sum_{n=0}^{\infty} p^n \cos(nx) = \Re\left[\sum_{n=0}^{\infty} p^n e^{inx}\right],$$

where $|p|<1$. The exponential sum formula gives,

$$\sum_{n=0}^{\infty} p^n \cos(nx) = \Re\left[\frac{1-pe^{-ix}}{1-2p\cos x + p^2}\right]$$
$$= \frac{1-p\cos x}{1-2p\cos x+p^2}.$$

The sum of $\cos^2(kx)$ can also be done in closed form,

$$\sum_{k=0}^{N} \cos^2(kx) = \frac{1}{4}\left\{3+2N+\csc x \sin\left[x(1+2N)\right]\right\}.$$

The Fourier transform of $\cos(2\pi k_0 x)$ is given by,

$$F_x[\cos(2\pi k_0 x)](k) = \int_{-\infty}^{\infty} e^{-2\pi i k x} \cos(2\pi k_0 x)\,dx$$
$$= \frac{1}{2}\left[\delta(k-k_0)+\delta(k+k_0)\right],$$

where $\delta(k)$ is the delta function.

Cvijović and Klinowski note that the following series,

$$C_v(\alpha) = \sum_{k=0}^{\infty} \frac{\cos(2k+1)\alpha}{(2k+1)^v}$$

has closed form for $v=2n$,

$$C_{2n}(\alpha) = \frac{(-1)^n}{4(2n-1)!}\pi^{2n} E_{2n-1}\left(\frac{\alpha}{\pi}\right),$$

where $E_n(x)$ is an Euler polynomial.

A definite integral involving $\cos x$ is given by,

$$\int_0^{\infty} \cos(x^n)\,dx = \Gamma\left(1+\frac{1}{n}\right)\cos\left(\frac{\pi}{2n}\right)$$

for $n>1$ where $\Gamma(z)$ is the gamma function.

Cosine Function is Even

Theorem:

For all $z \in \mathbb{C}$:

$$\cos(-z) = \cos z$$

That is, the cosine function is even.

Proof:

From the definition of the cosine function:

$$\cos z = \sum_{n=0}^{\infty} (-1)^n \frac{z^{2n}}{(2n)!} = 1 - \frac{z^2}{2!} + \frac{z^4}{4!} - \cdots$$

From Even Power is Non-Negative:

$$\forall n \in N : z^{2n} = (-z)^{2n}$$

The result follows.

Proof:

$$\cos(-z) = \frac{e^{i(-z)} + e^{-i(-z)}}{2} \qquad \text{Cosine Exponential Formulation}$$

$$= \frac{e^{iz} + e^{-iz}}{2} \qquad \text{simplifying}$$

$$= \cos z \qquad \text{Cosine Exponential Formulation}$$

Derivative of Cosine Function

Theorem:

$$D_x(\cos x) = -\sin x$$

Corollary:

$$D_x(\cos ax) = -a \sin ax$$

Proof:

From the definition of the cosine function, we have:

$$\cos x = \sum_{n=0}^{\infty} (-1)^n \frac{x^{2n}}{(2n)!}$$

Then:

$$D_x(\cos x) = \sum_{n=1}^{\infty} (-1)^n 2n \frac{x^{2n-1}}{(2n)!} \qquad \text{Power Series is Differentiable on Interval of Convergence}$$

$$= \sum_{n=1}^{\infty} (-1)^n \frac{x^{2n-1}}{(2n-1)!}$$

$$=\sum^{\infty}(-1)\ \frac{2\quad 1}{(2\quad 1)!}$$ changing summation index

$$=\sum^{\infty}(-\ )\ \frac{2\quad 1}{(2\quad 1)}$$

The result follows from the definition of the sine function.

Proof:

$$D_x(\cos x)=\lim_{h\to 0}\frac{\cos(x+h)-\cos(x)}{h}$$ Definition of Derivative of Real Function at Point

$$=\lim_{h\to 0}\frac{\cos(x)\cos(h)-\sin(x)\sin(h)-\cos(x)}{h}$$ Cosine of Sum

$$=\lim_{h\to 0}\frac{\cos(x)\cos(h)-\cos(x)}{h}+\lim_{h\to 0}\frac{-\sin(x)\sin(h)}{h}$$ Sum Rule for Limits of Functions

$$=\cos(x)\lim_{h\to 0}\frac{\cos(h)-1}{h}-\sin(x)\lim_{h\to 0}\frac{\sin(h)}{h}$$ Multiple Rule for Limits of Functions

$$=\cos(x)\times 0-\sin(x)\times 1$$

$$=-\sin(x)$$ Limit of (Cosine (X) - 1) over X and Limit of Sine of X over X

Proof:

$$D_x\cos x=D_x\sin\left(\frac{\pi}{2}-x\right)$$ Sine of Complement equals Cosine

$$=-\cos\left(\frac{\pi}{2}-x\right)$$ Derivative of Sine Function and Chain Rule

$$=-\sin x$$ Cosine of Complement equals Sine

Proof:

$$D_x(\cos x)=\lim_{h\to 0}\frac{\cos(x+h)-\cos(x)}{h}$$ Definition of Derivative of Real Function at Point

$$=\lim_{h\to 0}\frac{\cos\left(\left(x+\frac{h}{2}\right)+\frac{h}{2}\right)-\cos\left(\left(x+\frac{h}{2}\right)-\frac{h}{2}\right)}{h}$$

$$= \lim_{h \to 0} \frac{-2\sin\left(x + \frac{h}{2}\right)\sin\left(\frac{h}{2}\right)}{h}$$ Simpson's Formula for Sine by Sine

$$= \lim_{h \to 0} \sin\left(x + \frac{h}{2}\right) \lim_{h \to 0} \frac{\sin\left(\frac{h}{2}\right)}{\frac{h}{2}}$$ Multiple Rule for Limits of Functions and Product Rule for Limits of Functions

$$= -\sin(x) \times 1$$ Continuity of Sine and Limit of Sine of X over X

$$= -\sin(x)$$

Power Series Expansion for Cosine Function

Theorem:

The cosine function has the power series expansion:

$$\cos x = \sum_{n=0}^{\infty} (-1)^n \frac{x^{2n}}{(2n)!}$$
$$= 1 - \frac{x^2}{2!} + \frac{x^4}{4!} - \frac{x^6}{6!} + \dots$$

valid for all $x \in \mathbb{R}$.

Proof:

From Derivative of Cosine Function:

$$\frac{d}{dx}\cos x = -\sin x$$

From Derivative of Sine Function:

$$\frac{d}{dx}\sin x = \cos x.$$

Hence:

$$\frac{d^2}{dx^2}\cos x = -\cos x$$
$$\frac{d^3}{dx^3}\cos x = \sin x$$
$$\frac{d^4}{dx^4}\cos x = \cos x$$

and so for all $m \in \mathbb{N}$.

$$m = 4k: \quad \frac{d^m}{dx^m}\cos x = \cos x$$
$$m = 4k+1: \quad \frac{d^m}{dx^m}\cos x = -\sin x$$
$$m = 4k+2: \quad \frac{d^m}{dx^m}\cos x = -\cos x$$
$$m = 4k+3: \quad \frac{d^m}{dx^m}\cos x = \sin x$$

where $k \in \mathbb{Z}$.

This leads to the Maclaurin series expansion:

$$\sin x = \sum_{k=0}^{\infty}\left(\frac{x^{4k}}{(4k)!}\cos(0) - \frac{x^{4k+1}}{(4k+1)!}\sin(0) - \frac{x^{4k+2}}{(4k+2)!}\cos(0) + \frac{x^{4k+3}}{(4k+3)!}\sin(0)\right)$$

$$= \sum_{k=0}^{\infty}\left(\frac{x^{4k}}{(4k)!} - \frac{x^{4k+2}}{(4k+2)!}\right) \qquad \text{Sine of Zero is Zero, Cosine of Zero is One}$$

$$= \sum_{n=0}^{\infty}(-1)^n \frac{x^{2n}}{(2n)!} \qquad \text{setting } n = 2k$$

From Series of Power over Factorial Converges, it follows that this series is convergent for all x.

Complex Cosine Function

The complex function $\cos : \mathbb{C} \to \mathbb{C}$ is defined as:

$$\cos z = \sum_{n=0}^{\infty}(-1)^n \frac{z^{2n}}{(2n)!}$$
$$= 1 - \frac{z^2}{2!} + \frac{z^4}{4!} - \frac{z^6}{6!} + \cdots + (-1)^n \frac{z^{2n}}{(2n)!} + \cdots$$

Example: $4\cos z = 3 + i$

Let: $4\cos z = 3 + i$

Then:

$$z = \frac{(8n+1)\pi}{4} - \frac{i\ln 2}{2} \text{ for } n \in \mathbb{Z}$$

or:

$$z = \frac{(8m-1)i\pi}{4} + \frac{i\ln 2}{2} \text{ for } m \in \mathbb{Z}$$

Complex Cosine Function is Entire

Let $\cos : \mathbb{C} \to \mathbb{C}$ be the complex cosine function.

Then cos is entire.

Proof:

By the definition of the complex cosine function, cos admits a power series expansion about 0:

$$\cos z = \sum_{n=0}^{\infty} (-1)^n \frac{z^{2n}}{(2n)!}$$

By Complex Function is Entire iff it has Everywhere Convergent Power Series, to show that cos is entire it suffices to show that this series is everywhere convergent.

From Radius of Convergence from Limit of Sequence: Complex Case, it is sufficient to show that:

$$\lim_{n \to \infty} \left| \frac{(-1)^{n+1}}{(2n+2)!} \times \frac{(2n)!}{(-1)^n} \right| = 0$$

We have:

$$\begin{aligned} \lim_{n \to \infty} \left| \frac{(-1)^{n+1}}{(2n+2)!} \times \frac{(2n)!}{(-1)^m} \right| &= |-1| \lim_{n \to \infty} \left| \frac{(2n)!}{(2n+2)(2n+1)(2n)!} \right| && \text{Definition of Factorial} \\ &= \lim_{n \to \infty} \left(\frac{1}{(2n+2)(2n+1)} \right) \\ &= 0 \end{aligned}$$

hence the result.

Proof:

Let:

$$f(z) = \exp z$$
$$g(z) = iz$$
$$h(z) = -iz$$

for all $z \in \mathbb{C}$.

By Complex Exponential Function is Entire, we have that f is entire.

By Polynomial is Entire, we have that g and h are entire.

Therefore, by Composition of Entire Functions is Entire, we have that both f ∘ g and f ∘ h are entire.

By Linear Combination of Entire Functions is Entire, we then have that:

$$\frac{1}{2}(f \circ g + f \circ h)$$

is entire.

Note that:

$$\frac{1}{2}\big((f \circ g)(z) + (f \circ h)(z)\big) = \frac{1}{2}\big(\exp(iz) + \exp(-iz)\big)$$
$$= \cos z \qquad \text{Cosine Exponential Formulation}$$

Therefore, cos is an entire function.

Cosine Exponential Formulation

Theorem:

For any complex number $z \in \mathbb{C}$:

$$\cos z = \frac{\exp(iz) + \exp(-iz)}{2}$$

where:

exp z denotes the exponential function

cos z denotes the complex cosine function

i denotes the inaginary unit.

Real Domain

This result is often presented and proved separately for arguments in the real domain:

$$\cos x = \frac{e^{ix} + e^{-ix}}{2}$$

Proof:

From the definition of the cosine function:

$$\cos z = \sum_{n=0}^{\infty} (-1)^n \frac{z^{2n}}{(2n)!}$$

$$= 1 - \frac{z^2}{2!} + \frac{z^4}{4!} - \frac{z^6}{6!} + \cdots + (-1)^n \frac{z^{2n}}{(2n)!} + \cdots$$

From the definition of the exponential as a power series:

$$\exp z = \sum_{n=0}^{\infty} \frac{z^{2n}}{n!}$$

$$= 1 - \frac{z}{1!} + \frac{z^2}{2!} - \frac{z^3}{3!} + \cdots + \frac{z^n}{n!} + \cdots$$

Then, starting from the right hand side:

$$\frac{\exp(iz) + \exp(-iz)}{2} = \frac{1}{2}\left(\sum_{n=0}^{\infty} \frac{(iz)^n}{n!} + \sum_{n=0}^{\infty} \frac{(iz)^n}{n!} \right)$$

$$= \frac{1}{2} \sum_{n=0}^{\infty} \left(\frac{\left((iz)^n + (-iz)^n\right)}{n!} \right) \qquad \text{Cosine Function is Absolutely Convergent}$$

$$= \frac{1}{2} \sum_{n=0}^{\infty} \left(\frac{(iz)^{2n} + (-iz)^{2n}}{(2n)!} + \frac{(iz)^{2n+1} + (-iz)^{2n+1}}{(2n+1)!} \right) \qquad \text{split into even and odd } n$$

$$= \frac{1}{2} \sum_{n=0}^{\infty} \frac{(iz)^{2n} + (-iz)^{2n}}{(2n)!} \qquad (-iz)^{2n+1} = -(iz)^{2n+1}$$

$$= \frac{1}{2} \sum_{n=0}^{\infty} \frac{2(iz)^{2n}}{(2n)!} \qquad (-1)^{2n} = 1$$

$$= \sum_{n=0}^{\infty} \frac{(iz)^{2n}}{(2n)!} \qquad \text{cancel } 2$$

$$= \sum_{n=0}^{\infty} (-1)^n \frac{z^{2n}}{(2n)!} \qquad i^{2n} = (-1)^n$$

$$= \cos z$$

Proof:

From Euler's Formula:

$$\exp(iz) = \cos z + i \sin z$$

Then, starting from the right hand side:

$$\frac{\exp(iz) + \exp(-iz)}{2} = \frac{\cos z + i\sin z + \cos(-z) + i\sin(-z)}{2}$$

$$= \frac{\cos z + \cos(-z)}{2} \qquad \text{Sine Function is Odd}$$

$$= \frac{2\cos z}{2} \qquad \text{Cosine Function is Even}$$

$$= \cos z$$

Proof:

$$\exp(iz) = \cos z + i \sin z \qquad \text{Euler's Formula}$$

$$\exp(-iz) = \cos z - i \sin z \qquad \text{Euler's Formula : Corollary}$$

$$\leadsto \exp(iz) + \exp(-iz) = (\cos z + i \sin z) + (\cos z - i \sin z) \qquad (1) + (2)$$

$$= 2\cos z \qquad \text{simplifying}$$

$$\leadsto \frac{\exp(iz) + \exp(-iz)}{2} = \cos z$$

Also presented as:

$$\cos z = \frac{1}{2}\left(e^{-iz} + e^{iz}\right)$$

Laplace Transform of Cosine

Theorem:

Let cos be the real cosine function.

Let L{f} denote the Laplace transform of the real function f.

Then:

$$L\ \{\cos at\} = \frac{s}{s^2 + a^2}$$

where $a \in \mathbb{R}_{>0}$ is constant, and $\Re\mathfrak{e}(s) > a.$

Proof:

$$L\ \{\cos at\}(s) = \int_0^{\to +\infty} e^{-st} \cos at\, dt \qquad \text{Definition of Laplace Transform}$$

$$= \lim_{L\to\infty} \int_0^L e^{-st} \cos at\, dt \qquad \text{Definition of Improper Integral}$$

$$= \lim_{L\to\infty} \left[\frac{e^{-st}\left(-s\cos at + a\sin at\right)}{(-s)^2 + a^2} \right]^L \qquad \text{Primitive of } e^{ax}\cos bx$$

$$= \lim_{L\to\infty} \left(\frac{e^{-st}\left(-s\cos aL + a\sin aL\right)}{s^2 + a^2} - \frac{e^{-s\times 0}\left(-s\cos(0\times a) + a\sin(0\times a)\right)}{s^2 + a^2} \right)$$

$$= \lim_{L\to\infty} \left(\frac{s\cos(0\times a) - a\sin(0\times a)}{s^2 + a^2} - \frac{e^{-sL}\left(-s\cos aL + a\sin aL\right)}{s^2 + a^2} \right)$$

$$= \frac{s\cos(0\times a) - a\sin(0\times a)}{s^2 + a^2} - 0 \qquad \text{Exponential Tends to Zero}$$

$$= \frac{s\cos 0 - a\sin 0}{s^2 + a^2} \qquad \text{simplifying}$$

$$= \frac{s}{s^2 + a^2} \qquad \text{Sine of Zero is Zero, Cosine of Zero is One}$$

Proof:

$$L\ \{e^{iat}\} = \frac{1}{s - ia} \qquad \text{Laplace Transform of Exponential}$$

$$= \frac{s + ia}{s^2 + a^2} \qquad \text{multiply top and bottom by } s + ia$$

Also:

$$L\ \{e^{iat}\} = L\ \{\cos at + i\sin at\} \qquad \text{Euler's Formula}$$

$$= L\ \{\cos at\} + iL\ \{\sin at\} \qquad \text{Linear Combination of Laplace Transforms}$$

So:

$$\begin{aligned} L\ \{\cos at\} &= \mathfrak{Re}(L\ \{e^{iat}\}) \\ &= \mathfrak{Re}\left(\frac{s+ia}{s^2+a^2}\right) \\ &= \frac{s}{s^2+a^2} \end{aligned}$$

Proof:

$$\begin{aligned} L\ \{\cos at\} &= L\ \left\{\frac{e^{iat}+e^{-iat}}{2}\right\} && \text{Cosine Exponential Formulation} \\ &= \frac{1}{2}\left(L\ \left\{e^{iat}\right\}+L\ \left\{e^{-iat}\right\}\right) && \text{Linear Combination of Laplace Transforms} \\ &= \frac{1}{2}\left(\frac{1}{s-ia}+\frac{1}{s+ia}\right) && \text{Laplace Transform of Exponential} \\ &= \frac{1}{2}\left(\frac{s+ia+s-ia}{s^2+a^2}\right) && \text{simplifying} \\ &= \frac{s}{s^2+a^2} && \text{simplifying} \end{aligned}$$

Proof:

By definition of the Laplace Transform:

$$L\ \{\cos at\} = \int_0^{\to +\infty} e^{-st}\cos at\, dt$$

From Integration by Parts:

$$\int fg'\,dt = fg - \int f'g\,dt$$

Here:

$$\begin{aligned} & f = \cos at \\ \leadsto\quad & f' = -a\sin at && \text{Derivative of Cosine Function} \\ & g' = e^{-st} \\ \leadsto\quad & g = -\frac{1}{s}e^{-st} && \text{Primitive of Exponential Function} \end{aligned}$$

So:

$$\int e^{-st}\cos at\,dt = -\frac{1}{s}e^{-st}\cos at - \frac{a}{s}\int e^{-st}\sin at\,dt$$

Consider:

$$\int e^{-st}\sin at\,dt$$

Again, using Integration by Parts:

$$\int hj'\,\mathrm{dt} = hj - \int h'j\,\mathrm{dt}$$

Here:

$$h = \sin at$$

$$\leadsto \quad h' = a\cos at \qquad \text{Derivative of sine Function}$$

$$j' = e^{-st}$$

$$\leadsto \quad j = -\frac{1}{s}e^{-st} \qquad \text{Primitive of Exponential Function}$$

So:

$$\int e^{-st}\sin at\,dt = -\frac{1}{s}e^{-st}\sin at + \frac{a}{s}\int e^{-st}\cos at\,dt$$

Substituting this into equation: $\int e^{-st}\cos at\,dt = -\frac{1}{s}e^{-st}\cos at - \frac{a}{s}\int e^{-st}\sin at\,dt$

$$\int e^{-st}\cos at\,dt = -\frac{1}{s}e^{-st}\cos at - \frac{a}{s}\left(-\frac{1}{s}e^{-st}\sin at + \frac{a}{s}\int e^{-st}\cos at\,dt\right)$$

$$= -\frac{1}{s}e^{-st}\cos at + \frac{a}{s^2}e^{-st}\sin a - \frac{a^2}{s^2}\int e^{-st}\cos at\,dt$$

$$\leadsto \quad \left(1 + \frac{a^2}{s^2}\right)\int e^{-st}\cos at\,dt = -\frac{1}{s}e^{-st}\cos at + \frac{a}{s^2}e^{-st}\sin at$$

Evaluating at $t = 0$ and $t \to +\infty$:

$$\left(1 + \frac{a^2}{s^2}\right)L\,\{\cos at\} = \left[-e^{-st}\left(\frac{1}{s}\cos at - \frac{a}{s^2}\sin at\right)\right]_{t=0}^{t\to+\infty}$$

$$= 0 - \left(-1\left(\frac{1}{s}\times 1 + \frac{a}{s^2}\times 0\right)\right)$$

Boundedness of Real Sine and Cosine, Complex Exponential Tends to Zero

$$= \frac{1}{s}$$

$$\rightsquigarrow \qquad L\ \{\cos at\} = \frac{1}{s}\left(1 + \frac{a^2}{s^2}\right)^{-1}$$

$$= \frac{1}{s}\left(\frac{s^2}{a^2 + s^2}\right)$$

$$= \frac{s}{s^2 + a^2}$$

Tangent

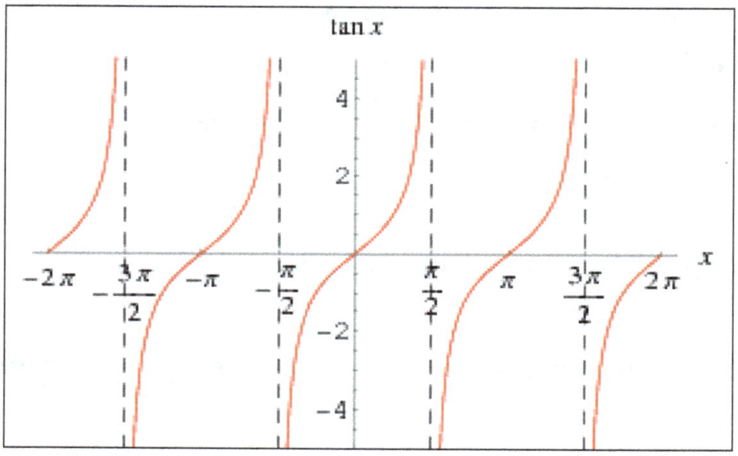

The tangent function is defined by,

$$\tan x \equiv \frac{\sin x}{\cos x},$$

where sin x is the sine function and cos x is the cosine function. The notation tg x is sometimes also used.

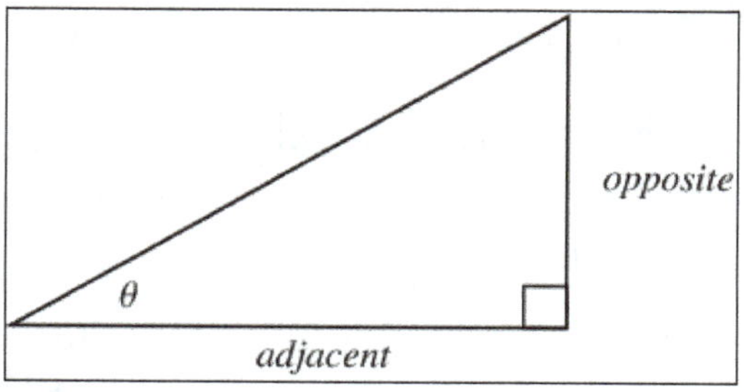

The common schoolbook definition of the tangent of an angle theta in a right triangle (which is equivalent to the definition just given) is as the ratio of the side lengths opposite to the angle and adjacent the angle, i.e.

$$\tan\theta \equiv \frac{\text{opposite}}{\text{adjacent}}.$$

A convenient mnemonic for remembering the definition of the sine, cosine, and tangent is SOHCAHTOA (sine equals opposite over hypotenuse, cosine equals adjacent over hypotenuse, tangent equals opposite over adjacent).

The word "tangent" also has an important related meaning as a line or plane which touches a given curve or solid at a single point. These geometrical objects are then called a tangent line or tangent plane, respectively.

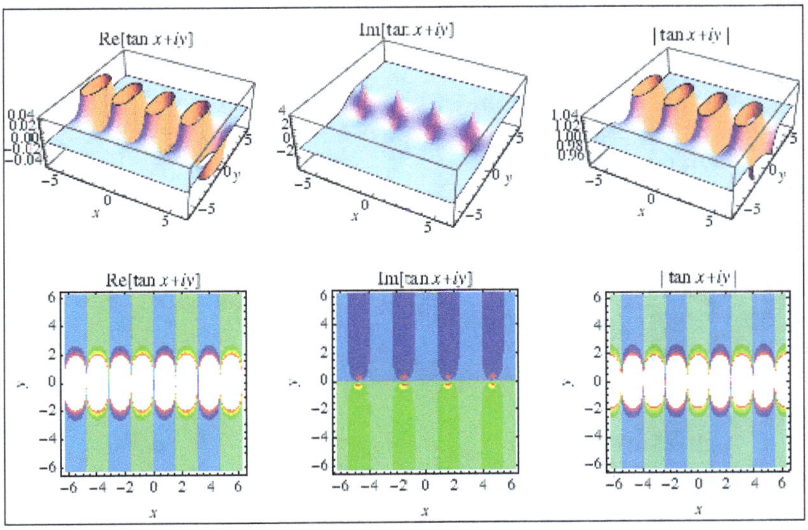

The definition of the tangent function can be extended to complex arguments z using the definition:

$$\begin{aligned}\tan z &= \frac{i(e^{-iz}-e^{iz})}{e^{-iz}+e^{iz}} \\ &= \frac{e^{iz}-e^{-iz}}{i\left(e^{iz}+e^{-iz}\right)} \\ &= \frac{i\left(1-e^{2iz}\right)}{1+e^{2iz}} \\ &= \frac{e^{2iz}-1}{i\left(e^{2iz}+1\right)},\end{aligned}$$

where e is the base of the natural logarithm and i is the imaginary number.

A related function known as the hyperbolic tangent is similarly defined,

$$\tan z = \frac{e^{z} - e^{-z}}{e^{z} + e^{-z}}.$$

An important tangent identity is given by:

$$\tan^2 \theta + 1 = \sec^2 \theta.$$

Angle addition, subtraction, half-angle, and multiple-angle formulas are given by:

$$\tan(\alpha + \beta) = \frac{\tan\alpha + \tan\beta}{1 - \tan\alpha \tan\beta}$$

$$\tan(\alpha - \beta) = \frac{\tan\alpha - \tan\beta}{1 + \tan\alpha \tan\beta}$$

$$\tan(2\alpha) = \frac{2\tan\alpha}{1 - \tan^2\alpha}$$

$$\tan(n\alpha) = \frac{\tan\left[(n-1)\alpha\right] + \tan\alpha}{1 - \tan\left[(n-1)\alpha\right]\tan\alpha}$$

$$\begin{aligned}\tan\left(\frac{\alpha}{2}\right) &= \frac{\sin\alpha}{1 + \cos\alpha} \\ &= \frac{1 - \cos\alpha}{\sin\alpha} \\ &= \frac{\tan\alpha \sin\alpha}{\tan\alpha + \sin\alpha}.\end{aligned}$$

The sine and cosine functions can conveniently be expressed in terms of a tangent as:

$$\cos t = \frac{1 - \tan^2\left(\frac{1}{2}t\right)}{1 + \tan^2\left(\frac{1}{2}t\right)}$$

$$\sin t = \frac{2\tan\left(\frac{1}{2}t\right)}{1 + \tan^2\left(\frac{1}{2}t\right)},$$

which can be particularly convenient in polynomial computations such as Gröbner basis since it reduces the number of equations compared with explicit inclusion of cos t and sin t together with the additional relation $\cos^2 t + \sin^2 t - 1 = 0$.

These lead to the pretty identity,

$$\tan\left(x+\frac{1}{4}\pi\right)=\frac{1+\tan x}{1-\tan x}.$$

There is also a beautiful angle addition identity for three variables,

$$\tan(\alpha+\beta+\gamma)=\frac{\tan\alpha+\tan\beta+\tan\gamma-\tan\alpha\tan\beta\tan\gamma}{1-\tan\beta\tan\gamma-\tan\gamma\tan\alpha-\tan\alpha\tan\beta}.$$

Another tangent identity is,

$$\begin{aligned}\tan(nx)&=\frac{\sum_{k=0}^{n}(-1)^k\binom{n}{2k+1}t^{2k+1}}{\sum_{k=0}^{n}(-1)^k\binom{n}{2k}t^{2k}}\\&=\frac{\frac{1}{2}i(1-it)^n-(1+it)^n}{\frac{1}{2}(1-it)^n+(1+it)^n}\\&=\frac{1}{i}\frac{(1-it)^n-(1+it)^n}{(1-it)^n+(1-it)^n},\end{aligned}$$

where t = tan x. Written explicitly,

$$\tan(nx)=\frac{2i\left(1-i\tan t^n\right)}{(1-i\tan t)^n+(1+i\tan t)^n}-\mathrm{i},$$

This gives the first few expansions as,

$$\tan x=t$$

$$\tan(2x)=\frac{2t}{1-t^2}$$

$$\tan(3x)=\frac{3t-t^3}{1-3t^2}$$

$$\tan(4x)=\frac{4t-4t^3}{1-6t^2+r^4}$$

$$\tan(5x)=\frac{5t-10t^3+t^5}{1-10t^2+5t^4}.$$

A beautiful formula that generalizes the tangent angle addition formula, equation $\tan(4x)=\frac{4t-4t^3}{1-6t^2+r^4}$ and $\tan(5x)=\frac{5t-10t^3+t^5}{1-10t^2+5t^4}$ is given by,

$$\tan\left(\sum_{n=1}^{N}\theta_n\right)=i\frac{\prod_{n=1}^{N}(1-i\tan\theta_n)-\prod_{n=1}^{N}(i\tan\theta_n+1)}{\prod_{n=1}^{N}(i\tan\theta_n+1)+\prod_{n=1}^{N}(1-i\tan\theta_n)}$$

There are a number of simple but interesting tangent identities based on those given above, including,

$$\tan(A+60°)\tan(A-60°)+\tan A\tan(A+60°)+\tan A\tan(A-60°)=-3.$$

The Maclaurin series valid for $-\pi/2<x<\pi/2$ for the tangent function is

$$\tan x=\sum_{n=1}^{\infty}\frac{(-1)^{n-1}2^{2n}\left(2^{2n-1}\right)B_{2n}}{(2n)!}x^{2n-1}$$
$$=x+\frac{1}{3}x^3+\frac{2}{15}x^5+\frac{17}{315}x^7+\frac{62}{2835}x^9+\ldots$$

where B_n is a Bernoulli number.

tan x is irrational for any rational $x\neq 0$, which can be proved by writing tan x as a continued fraction as,

$$\tan x\frac{x}{1-\frac{x^2}{3-\frac{x^2}{5-\frac{x^2}{7-\ldots}}}}$$

and,

$$\tan x=\frac{1}{\frac{1}{x}-\frac{1}{\frac{3}{x}-\frac{1}{\frac{5}{x}-\frac{1}{\frac{7}{x}-\ldots}}}}$$

both due to Lambert.

An interesting identity involving the product of tangents is,

$$\prod_{k=1}^{\lfloor (n-1)/2\rfloor}\tan\left(\frac{k\pi}{n}\right)=\begin{cases}\sqrt{n} & \text{for } n \text{ odd}\\ 1 & \text{for } n \text{ even}\end{cases}$$

where $\lfloor x\rfloor$ is the floor function.

The equation $x = \tan x$, which is equivalent to $\operatorname{tanc}(x) = 1$, where $\operatorname{tanc}(x)$ is the tanc function, does not have simple closed-form solutions.

The difference between consecutive solutions gets closer and closer to for higher order solutions. The function $\operatorname{tanc}(x) \equiv (\tan x)/x$ is sometimes known as the tanc function.

Tangent Function is Odd

Theorem:

For all $x \in \mathbb{C}$ where tan x is defined:

$$\tan(-x) = -\tan x$$

That is, the tangent function is odd.

Proof:

$$\begin{aligned} \tan(-x) &= \frac{\sin(-x)}{\cos(-x)} && \text{Tangent is Sine divided by Cosine} \\ &= \frac{-\sin x}{\cos x} && \text{Sine Function is Odd; Cosine Function is Even} \\ &= -\tan x && \text{Tangent is Sine divided by Cosine} \end{aligned}$$

Derivative of Tangent Function

Theorem:

$$D_x(\tan x) = \sec^2 x = \frac{1}{\cos^2 x}$$

when $\cos x \neq 0$.

Corollary:

$$D_x(\tan ax) = a\sec^2 ax$$

Proof:

From the definition of the tangent function:

$$\tan x = \frac{\sin x}{\cos x}$$

From Derivative of Sine Function:

$$D_x(\sin x) = \cos x$$

From Derivative of Cosine Function:

$$D_x(\cos x) = -\sin x$$

Then:

$$D_x(\tan x) = \frac{\cos x \cos x - \sin x(-\sin x)}{\cos^2 x} \quad \text{Quotient Rule for Derivatives}$$

$$= \frac{\cos^2 x + \sin^2 x}{\cos^2 x}$$

$$= \frac{1}{\cos^2 x} \quad \text{Sum of Squares of Sine and Cosine}$$

This is valid only when $\cos x \neq 0$.

The result follows from the Secant is Reciprocal of Cosine.

Proof:

$$D_x(\tan x) = \lim_{h\to 0} \frac{\tan(x+h) - \tan(x)}{h} \quad \text{Definition of Derivative of Real Function at Point}$$

$$= \lim_{h\to 0} \frac{\dfrac{\tan(x) + \tan(h)}{1 - \tan(x)\tan(h)} - \tan(x)}{h} \quad \text{Tangent of Sum}$$

$$= \lim_{h\to 0} \frac{\dfrac{\tan(x) + \tan(h) - \tan(x) + \tan^2(x)\tan(h)}{1 - \tan(x)\tan(h)}}{h}$$

$$= \lim_{h\to 0} \frac{\tan(h) + \tan^2(x)\tan(h)}{h(1 - \tan(x)\tan(h))}$$

$$= \lim_{h\to 0} \frac{1 + \tan^2(x)}{1 - \tan(x)\tan(h)} \cdot \lim_{h\to 0} \frac{\tan(h)}{h} \quad \text{Product Rule for Limits of Functions}$$

$$= \frac{1 + \tan^2 x}{1 - \tan(x)\tan(0)} \cdot 1 \quad \text{Limit of Tan X over X}$$

$$= 1 + \tan^2(x) \quad \text{Tangent of Zero}$$

$$= \sec^2(x)$$

Corollary to Sum of Squares of Sine and Cosine

$$= \frac{1}{\cos^2(x)}$$

Secant is Reciprocal of Cosine $\cos(x) \neq 0$

Power Series Expansion for Tangent Function

Theorem:

The tangent function has a Taylor series expansion:

$$\tan x = \sum_{n=1}^{\infty} \frac{(-1)^{n-1} 2^{2n} \left(2^{2n-1}\right) B_{2n}\, x^{2n-1}}{(2n)!}$$

$$= x + \frac{x^3}{3} + \frac{2x^5}{15} + \frac{17x^7}{315} \ldots$$

where B_{2n} denotes the Bernoulli numbers.

This converges for $|x| < \frac{\pi}{2}$.

Proof:

From Power Series Expansion for Cotangent Function:

$$\cot x = \sum_{n=0}^{\infty} \frac{(-1)^n 2^{2n} B_{2n}\, x^{2n-1}}{(2n)!}$$

Then:

$$\tan x = \cot x - 2\cot 2x$$

Cotangent Minus Tangent

$$= \sum_{n=0}^{\infty} \frac{(-1)^n 2^{2n} B_{2n}\, x^{2n-1}}{(2n)!} - 2\sum_{n=0}^{\infty} \frac{(-1)^n 2^{2n} B_{2n} (2x)^{2n-1}}{(2n)!}$$

$$= \sum_{n=0}^{\infty} \frac{(-1)^n 2^{2n} \left(1 - 2^{2n}\right) B_{2n}\, x^{2n-1}}{(2n)!}$$

$$= \sum_{n=1}^{\infty} \frac{(-1)^{n-1} 2^{2n} \left(2^{2n} - 1\right) B_{2n}\, x^{2n-1}}{(2n)!}$$

Proof:

We have:

$$\frac{x}{e^x - 1} = \frac{x}{2}\left(\frac{2}{e^x - 1}\right)$$

$$= \frac{x}{2}\left(\frac{e^x - e^x + 2}{e^x - 1}\right)$$

$$= \frac{x}{2}\left(\frac{\left(e^x + 1\right) - \left(e^x - 1\right)}{e^x - 1}\right)$$

$$= \frac{x}{2}\left(\frac{e^x + 1}{e^x - 1} - 1\right)$$

$$= -\frac{x}{2} + \frac{x}{2}\left(\frac{e^x + 1}{e^x - 1}\right)$$

$$= -\frac{x}{2} + \frac{x}{2}\left(\frac{e^{x/2} + e^{-x/2}}{e^{x/2} - e^{-x/2}}\right) \qquad \text{multiplying top and bottom by } e^{-x/2}$$

Thus:

$$\frac{x}{2}\left(\frac{e^{x/2} + e^{-x/2}}{e^{x/2} - e^{-x/2}}\right) = \sum_{n=0}^{\infty} \frac{B_{2n}}{(2n!)} x^{2n} \qquad \text{Definition of Bernoulli numbers}$$

Replacing x with 2ix in the left hand side equation above:

$$\frac{ix}{\quad}\left(\frac{e^{2ix/2} \quad e^{\ 2i\ \ /2}}{e^{2ix/2} \quad e^{\ 2ix/2}}\right)$$

$$ix\left(\frac{e^{ix} \quad e^{\ ix}}{e^{ix} \quad e}\right)$$

$$x \cot x \qquad \text{Cotangent Exponential Formulation}$$

Replacing x with 2ix in the right hand side in equation

$$\frac{x}{2}\left(\frac{e^{x/2} + e^{-x/2}}{e^{x/2} - e^{-x/2}}\right) = \sum_{n=0}^{\infty} \frac{B_{2n}}{(2n!)} x^{2n} \qquad \text{Definition of Bernoulli numbers}$$

$$= \sum_{n=0}^{\infty} \frac{B_{2n}}{(2n)!} (2ix)^{2n}$$

$$= \sum_{n=0}^{\infty} (-1)^n \frac{2^{2n} B_{2n}}{(2n)!} x^{2n}$$

$$\leadsto \quad x\cot x = \sum_{n=0}^{\infty} (-1)^n \frac{2^{2n} B_{2n}}{(2n)!} x^{2n}$$

$$\leadsto \quad \cot x = \sum_{n=0}^{\infty} (-1)^n \frac{2^{2n} B_{2n}}{(2n)!} x^{2n-1}$$

Then from Cotangent Minus Tangent:

$$\tan x = \cot x - 2\cot 2x$$

from which:

$$\begin{aligned}
\tan x &= \sum_{n=0}^{\infty} \frac{(-1)^n 2^{2n} B_{2n}}{(2n)!} x^{2n-1} - 2\sum_{n=0}^{\infty} \frac{(-1)^n 2^{2n} B_{2n}}{(2n)!} (2x)^{2x-1} \\
&= \sum_{n=0}^{\infty} \frac{(-1)^n 2^{2n} \left(1 - 2^{2n}\right) B_{2n}}{(2n)!} x^{2n-1} \\
&= \sum_{n=1}^{\infty} \frac{(-1)^{n-1} 2^{2n} \left(2^{2n} - 1\right) B_{2n}}{(2n)!} x^{2n-1}
\end{aligned}$$

Proof of Convergence

By Combination Theorem for Limits of Functions we can deduce the following:

$$\begin{aligned}
&\lim_{n\to\infty} \left| \frac{\dfrac{(-1)^n 2^{n+2} \left(2^{n+2} - 1\right) B_{2n+2}}{(2n+2)!} x^{2n+1}}{\dfrac{(-1)^{n-1} 2^{2n} \left(2^{2n} - 1\right) B_{2n}}{(2n)!} x^{2n-1}} \right| \\
&= \lim_{n\to\infty} \left| \frac{\left(2^{n+2} - 1\right)}{\left(2^{n+2} - 1\right)} \frac{1}{(2n+1)(2n+2)} \frac{B_{2n+2}}{B_{2n}} \right| 4x^2 \\
&= \lim_{n\to\infty} \left| \frac{2^{2n+2} - 1}{2^{2n} - 1} \right| \left| \frac{1}{(2n+1)(n+1)} \frac{B_{2n+2}}{B_{2n}} \right| 2x^2 \\
&= \lim_{n\to\infty} \left| 4 \frac{2^{2n}}{2^{2n} - 1} - \frac{1}{2^{2n} - 1} \right| \left| \frac{1}{(2n+1)(n+1)} \frac{B_{2n+2}}{B_{2n}} \right| 2x^2 \\
&= \lim_{n\to\infty} \left| 4 + \frac{4}{2^{2n} - 1} - \frac{1}{2^{2n} - 1} \right| \left| \frac{1}{(2n+1)(n+1)} \frac{B_{2n+2}}{B_{2n}} \right| 2x^2 \\
&= \lim_{n\to\infty} \left| \frac{1}{(2n+1)(n+1)} \frac{B_{2n+2}}{B_{2n}} \right| 8x^2
\end{aligned}$$

$$= \lim_{n\to\infty} \left| \frac{1}{(2n+1)(n+1)} \frac{(-1)^{n+2} 4\sqrt{\pi(n+1)}\left(\frac{n+1}{\pi e}\right)^{2n+2}}{(-1)^{n+1} 4\sqrt{\pi n}\left(\frac{n}{\pi e}\right)^{2n}} \right| 8x^2$$ Asymptotic Formula for Bernoulli Numbers

$$= \lim_{n\to\infty} \left| \frac{(n+1)^2}{(2n+1)(n+1)} \sqrt{\frac{n+1}{n}} \left(\frac{n+1}{n}\right)^{2n} \right| \frac{8}{\pi^2 e^2} x^2$$

$$= \lim_{n\to\infty} \left| \left(\frac{n+1}{n}\right)^{2n} \right| \frac{4}{\pi^2 e^2} x^2$$

$$= \lim_{n\to\infty} \left| \left(\left(1+\frac{1}{n}\right)^n\right)^2 \right| \frac{4}{\pi^2 e^2} x^2$$

$$= \frac{4}{\pi^2 e^2} x^2$$ Definition of Euler's Number

$$= \frac{4}{\pi^2} x^2$$

This is less than 1 if and only if:

$$|x| < \frac{\pi}{2}$$

Hence by the Ratio Test, the series converges for $|x| < \frac{\pi}{2}$.

Sequence of Terms

The Power Series Expansion for Tangent Function begins:

$$\tan x = x + \frac{1}{3}x^3 + \frac{2}{15}x^5 + \frac{17}{315}x^7 + \frac{62}{2835}x^9 + \cdots$$

Tangent Exponential Formulation

Theorem:

Let z be a complex number.

Let tan z denote the tangent function and i denote the imaginary unit: $i^2 = -1$.

Then.

Formulation:

$$\tan z = i\frac{1-e^{2iz}}{1+e^{2iz}}$$

Formulation:

$$\tan z = \frac{e^{iz}-e^{-iz}}{i\left(e^{iz}+e^{-iz}\right)}$$

Formulation:

$$\tan z = -i\left(\frac{e^{iz}-e^{-iz}}{e^{iz}+e^{-iz}}\right).$$

Cosecant

The cosecant function is an old mathematical function. It was mentioned in the works of G. J. von Lauchen Rheticus and E. Gunter. It was widely used by L. Euler and T. Olivier, Wait, and Jones.

The classical definition of the cosecant function for real arguments is: "the cosecant of an angle in a right-angle triangle is the ratio of the length of the hypotenuse to the length of the opposite leg." This description of $\csc(\alpha)$ is valid for $0<\alpha<\pi/2$ when this triangle is nondegenerate. This approach to the cosecant can be expanded to arbitrary real values of α if the arbitrary point (x,y) in the x,y-Cartesian plane is considered and $\csc(\alpha)$ is defined as the ratio $\left(x^2+y^2\right)^{1/2}/y$ assuming that α is the value of the angle between the positive direction of the x-axis and the direction from the origin to the point (x,y).

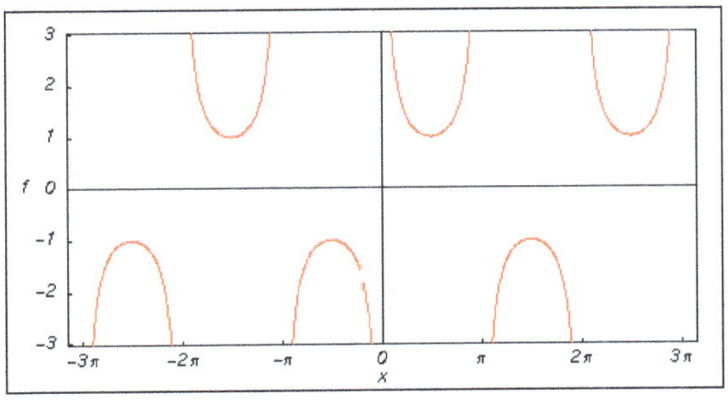

Comparing the classical definition with the definition of the sine function shows that the following formula can also be used as a definition of the cosecant function:

$$\csc(z) = \frac{1}{\sin(z)}.$$

Here is a graphic of the cosecant function $f(x)=\csc(x)$ for real values of its argument x.

Representation through more General Functions

The cosecant function $\csc(z)$ can be represented using more general mathematical functions. As the ratio of one divided by the sine function that is a particular case of the generalized hypergeometric, Bessel, Struve, and Mathieu functions, the cosecant function can also be represented as ratios of one and those special functions. Here are some examples:

$$\csc(z)=\frac{1}{z_0F_1\left(;\frac{3}{2};-\frac{z^2}{4}\right)}$$

$$\csc(z)=\sqrt{\frac{2}{\pi}}\frac{1}{\sqrt{z}\,J_{\frac{1}{2}}(z)}$$

$$\csc(z)=\sqrt{\frac{2}{\pi}}\frac{1}{\sqrt{z}\,I_{\frac{1}{2}}(iz)}$$

$$\csc(z)=\sqrt{\frac{2}{\pi}}\frac{1}{\sqrt{z}\,Y_{-\frac{1}{2}}(z)}$$

$$\csc(z)=\frac{\sqrt{\frac{2}{\pi}}}{\sqrt{z}\,H_{-\frac{1}{2}}(z)}$$

$$\csc(z)=\frac{1}{Se(1,0,z)}.$$

But these representations are not very useful because they include complicated special functions in the denominators.

It is more useful to write the cosecant function as particular cases of one special function. That can be done using doubly periodic Jacobi elliptic functions that degenerate into the cosecant function when their second parameter is equal to 0 or 1:

$$\csc(z)=ds(z|0)=ns(z|0)=dc\left(\frac{\pi}{2}-z|0\right)=$$
$$\mathrm{nc}\left(\frac{\pi}{2}-z|0\right)=i\,\mathrm{cs}(iz|1)=i\,\mathrm{ds}(iz|1)=\mathrm{cn}\left(\frac{\pi i}{2}-iz\right)=\mathrm{dn}\left(\frac{\pi i}{2}-iz|1\right).$$

Definition of the Cosecant Function for a Complex Argument

In the complex z-plane, the function $\csc(z)$ is defined using $\sin(z)$ or the exponential function e^w in the points iz and $-iz$ through the formula:

$$\csc(z) = \frac{1}{\sin(z)} = \frac{2i}{e^{iz} - e^{-iz}}.$$

In the points $z = \pi k /; k \in \mathbb{Z}$, where $\sin(z)$ is zero, the denominator of the last formula equals zero and $\csc(z)$ has singularities (poles of the first order).

Here are two graphics showing the real and imaginary parts of the cosecant function over the complex plane.

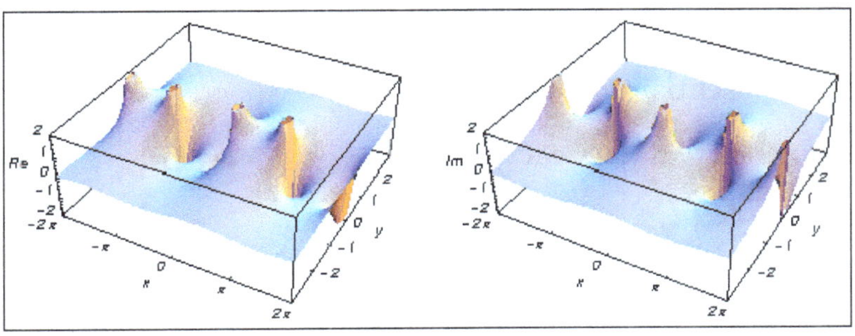

The best-known properties and formulas for the cosecant function:

Values in Points

Using the connection between the sine and cosecant functions, the following table of cosecant function values for angles between 0 and 2π can be derived:

$$\csc(0) = \overline{\infty} \qquad \csc\left(\frac{\pi}{6}\right) = 2 \qquad \csc = \left(\frac{\pi}{4}\right) = \sqrt{2} \qquad \csc\left(\frac{\pi}{3}\right) = \frac{2}{\sqrt{3}}$$

$$\csc\left(\frac{\pi}{2}\right) = 1 \qquad \csc\left(\frac{2\pi}{3}\right) = \frac{2}{\sqrt{3}} \qquad \csc\left(\frac{3\pi}{4}\right) = \sqrt{2} \qquad \csc\left(\frac{5\pi}{6}\right) = 2$$

$$\csc(\pi) = \overline{\infty} \qquad \csc\left(\frac{7\pi}{6}\right) = -2 \qquad \csc\left(\frac{5\pi}{4}\right) = -\sqrt{2} \qquad \csc\left(\frac{4\pi}{3}\right) = -\frac{2}{\sqrt{3}}$$

$$\csc\left(\frac{3\pi}{2}\right) = -1 \qquad \csc\left(\frac{5\pi}{3}\right) = -\frac{2}{\sqrt{3}} \qquad \csc\left(\frac{7\pi}{4}\right) = -\sqrt{2} \qquad \csc\left(\frac{11\pi}{6}\right) = -2$$

$$\csc(2\pi) = \overline{\infty} \qquad \csc(\pi m) = \overline{\infty} /; m \in \mathbb{Z} \qquad \csc\left(\pi\left(\frac{1}{2} + m\right)\right) = (-1)^m /; m \in \mathbb{Z}.$$

For real values of argument z, the values of $\csc(z)$ are real.

In the points $z = 2\pi n / m /; n \in \mathbb{Z} \wedge m \in \mathbb{Z}$, the values of $\csc(z)$ are algebraic. In several cases they can be integers -2, -1, -1, or 2:

$$\csc\left(-\frac{\pi}{2}\right) = -1 \quad \csc\left(-\frac{\pi}{6}\right) = -2 \quad \csc\left(\frac{\pi}{2}\right) = 1 \quad \csc\left(\frac{\pi}{6}\right) = 2.$$

The values of $\csc\left(\frac{n\pi}{m}\right)$ can be expressed using only square roots if $n \in \mathbb{Z}$ and m is a product of a power of 2 and distinct Fermat primes {3, 5, 17, 257, ...}.

The function $\csc(z)$ is an analytical function of z that is defined over the whole complex z-plane and does not have branch cuts and branch points. It has an infinite set of singular points:

- $z = \pi k /; k \in \mathbb{Z}$ are the simple poles with residues $(-1)^k$.
- $z = \tilde{\infty}$ is an essential singular point.

It is a periodic function with the real period 2π:

$$\csc(z + 2\pi) = \csc(z)$$

$$\csc(z) = \csc(z + 2\pi k) /; k \in \mathbb{Z} \quad \csc(z) = (-1)^k \csc(z + \pi k) /; k \in \mathbb{Z}.$$

The function $\csc(z)$ is an odd function with mirror symmetry:

$$\csc(-z) = -\csc(z) \quad \csc(\bar{z}) = \overline{\csc(z)}.$$

Differentiation

The first derivative of $\csc(z)$ has simple representations using either the $\cot(z)$ function or the $\csc(z)$ function:

$$\frac{\partial \csc(z)}{\partial z} = -\cot(z)\csc(z).$$

The nth derivative of $\csc(z)$ has much more complicated representations than symbolic n[th] derivatives for $\sin(z)$ and $\csc(z)$:

$$\frac{\partial \csc(z)}{\partial z^n} = \cot(z)\left(\delta_n + (n+1)! \sum_{k=0}^{n} \sum_{j=0}^{\left[\frac{k-1}{2}\right]} \frac{(-1)^{j+n} 2^{1-k} (k-2j)^n \csc^k(z)}{(k+1)\, j!(k-j)!(n-k)!} \cos\left(\frac{\pi(k-n)}{2} + (k-2j)z\right)\right) /; n \in \mathbb{N},$$

Where δ_n is the Kronecker delta symbol: $\delta_0 = 1$ and $\delta_n = 0 /; n \neq 0$.

Ordinary Differential Equation

The function $\csc(z)$ satisfies the following first-order nonlinear differential equation:

$$w'(z)^2 - w(z)^4 + w(z)^2 = 0 /; w(z) = \csc(z).$$

Series Representation

The function $\csc(z)$ has the following Laurent series expansion at the origin that converges for all finite values z with $0 < |z| < \pi$:

$$\csc(z) = \frac{1}{z} + \frac{z}{6} + \frac{7z^2}{360} + \ldots = \frac{1}{z} + \sum_{k=1}^{\infty} \frac{(-1)^{k-1} 2(2^{2k-1} - 1) B_{2k} z^{2k-1}}{(2k)!},$$

where B_{2k} are the Bernoulli numbers.

The cosecant function $\cos(z)$ can also be represented using other kinds of series by the following formulas:

$$\csc(z) = \frac{1}{z} + 2z \sum_{k=1}^{\infty} \frac{(-1)^k}{z^2 - \pi^2 k^2} /; \frac{z}{\pi} \notin \mathbb{Z}$$

$$\csc^2(z) = \sum_{k=-\infty}^{\infty} \frac{1}{(z - \pi k^2)} /; \frac{z}{\pi} \notin \mathbb{Z}.$$

Integral Representation

The function $\csc(z)$ has well-known integral representation through the following definite integral along the positive part of the real axis:

$$\csc(z) = \frac{1}{\pi} \int_0^{\infty} \frac{1}{t^2 + t} t^{z/\pi} \, dt /; 0 < \operatorname{Re}(z) < \pi$$

Product Representation

The famous infinite product representation for $\sin(z)$ can be easily rewritten as the following product representation for the cosecant function:

$$\csc(z) = \frac{1}{z} \prod_{k=1}^{\infty} \frac{\pi^2 k^2}{\pi^2 k^2 - z^2}.$$

Limit Representation

The cosecant function has the following limit representation:

$$\csc(z) = \lim_{n\to\infty} \sum_{k=-n}^{n} \frac{(-1)^k}{z - \pi k} /; \frac{z}{\pi} \notin \mathbb{Z}.$$

Indefinite Integration

Indefinite integrals of expressions that contain the cosecant function can sometimes be expressed using elementary functions. However, special functions are frequently needed to express the results even when the integrands have a simple form (if they can be evaluated in closed form). Here are some examples:

$$\int \csc(z)\,dz = \log\left(\tan\left(\frac{z}{z}\right)\right)$$

$$\int \sqrt{\csc(z)}\,dz = -2\csc^{\frac{1}{2}}(z)\sin^{\frac{1}{2}}(z)\,F\left(\frac{1}{4}(\pi - 2z)\middle|2\right)$$

$$\int \csc^{\nu}(a z)\,dz = -\frac{\cos(a z)\csc^{\nu-1}(a z)\sin^2(a z)^{\frac{\nu-1}{2}}}{a}\,{}_2F_1\left(\frac{1}{2},\frac{\nu+1}{2};\frac{3}{2};\cos^2(a z)\right).$$

Definite Integration

Definite integrals that contain the cosecant function are sometimes simple. For example, the famous Catalan constant C can be defined as the value of the following integral:

$$\frac{1}{2}\int_0^{\pi/2} t\csc(t)\,dt = c.$$

This constant also appears in the following integral:

$$\int_0^{\frac{\pi}{2}} t^2\csc(t)\,dt = 2C\pi - \frac{7\zeta(3)}{2}.$$

Some special functions can be used to evaluate more complicated definite integrals. For example, polylogarithmical, zeta, and gamma functions are needed to express the following integrals:

$$\int_0^{\frac{\pi}{3}} t^3\csc^2(t)\,dt =$$
$$\frac{1}{162}\left(6i\pi^3 - 2\sqrt{3}\pi^3 + 54\pi^2\log\left(-\sqrt[3]{-1}\left(-1+(-1)^{2/3}\right)\right) + 162i\pi\,\mathrm{Li}_2\left(-\sqrt[3]{-1}\right) + 243\,\mathrm{Li}_3\left(-\sqrt[3]{-1}\right) - 243\zeta(3)\right)$$

$$\int_0^{\frac{\pi}{2}} \csc^a(t)\,dt = \frac{\sqrt{\pi}}{2\Gamma\left(1-\frac{a}{2}\right)}\Gamma\left(1-\frac{a}{2}\right) /; \mathrm{Re}(a)<1.$$

Finite Summation

The following finite sums that contain the cosecant function have simple values:

$$\sum_{k=0}^{n-1}\csc^2\left(\frac{\pi k}{n}+z\right)=n^2\csc^2(nz)/;n\in\mathbb{N}^-$$

$$\sum_{k=1}^{n-1}\csc^2\left(\frac{k\pi}{n}\right)=\frac{1}{3}\left(n^2-1\right)/;n\in\mathbb{N}^+$$

$$\sum_{k=1}^{n}\csc^2\left(\frac{k\pi}{2n+1}\right)=\frac{2}{3}n(n+1)/;n\in\mathbb{N}^+$$

$$\sum_{k=1}^{\left\lfloor\frac{n-1}{2}\right\rfloor}\csc^2\left(\frac{k\pi}{n}\right)=\frac{1}{12}\left(2n^2-3(-1)^n-5\right)/;n\in\mathbb{N}^+$$

$$\sum_{k=0}^{\left\lfloor\frac{n-2}{2}\right\rfloor}\csc^2\left(\frac{(2k+1)\pi}{2n}\right)=\frac{1}{4}\left(2n^2+(-1)^n-1\right)/;n\in\mathbb{N}^+$$

$$\sum_{k=0}^{n}\csc\left(\frac{z}{2^k}\right)=\cot\left(\frac{z}{2^{n+1}}\right)-\cot(z)/;n\in\mathbb{N}^+.$$

Infinite Summation

The following infinite sum that contains the cosecant has a simple value:

$$\sum_{k=1}^{\infty}\frac{\csc k\pi\sqrt{2}}{k^3}=\frac{13\pi^3}{360\sqrt{2}}.$$

Finite Products

The following finite product from the cosecant can also be represented using the cosecant function:

$$\prod_{k=0}^{n-1}\csc\left(z+\frac{\pi k}{n}\right)=2^{n-1}\csc(nz)/;n\in\mathbb{N}^+.$$

Addition Formulas

The cosecant of a sum and the cosecant of a difference can be represented by the formulas that follow from corresponding formulas for the sine of a sum and the sine of a difference:

$$\csc(a+b)=\frac{1}{\cos(b)\sin(a)+\cos(a)\sin(b)}$$

$$\csc(a-b)=\frac{1}{\cos(b)\sin(a)-\cos(a)\sin(b)}.$$

Multiple Arguments

In the case of multiple arguments z, 2z, 3z, ..., the function $\csc(z)$ can be represented as a rational function that contains powers of cosecants and secants. Here are two examples:

$$\csc(2z) = \frac{1}{2}\csc(z)\sec(z)$$

$$\csc(3z) = \frac{\csc^3(z)}{3\csc^2(z) - 4}.$$

Half-angle Formulas

The cosecant of a half-angle can be represented by the following simple formula that is valid in a vertical strip:

$$\csc\left(\frac{z}{2}\right) = \frac{\sqrt{2}}{\sqrt{1-\cos(z)}} /; 0 < \mathrm{Re}(z) < 2\pi \vee \mathrm{Re}(z) = 0 \wedge \mathrm{Im}(z) > 0 \vee \mathrm{Re}(z) = 2\pi \wedge \mathrm{Im}(z) < 0.$$

To make this formula correct for all complex z, a complicated prefactor is needed:

$$\csc\left(\frac{z}{2}\right) = \frac{c(z)}{\sqrt{1-\cos(z)}} /; \mathrm{c}(z) = \sqrt{2}(-1)^{\left\lfloor \frac{\mathrm{Re}(z)}{2\pi} \right\rfloor}\left(1 - \left(1 + (-1)^{\left\lfloor \frac{\mathrm{Re}(z)}{2\pi} \right\rfloor + \left\lfloor -\frac{\mathrm{Re}(z)}{2\pi} \right\rfloor}\right)\theta(-\mathrm{Im}(z))\right),$$

where $\mathrm{c}(z)$ contains the unit step, real part, imaginary part, and the floor functions.

Sums of two Direct Functions

The sum and difference of two cosecant functions can be described by the following formulas:

$$\csc(a) + \csc(b) = 2\sin\left(\frac{a}{2} + \frac{b}{2}\right)\cos\left(\frac{a}{2} - \frac{b}{2}\right)\csc(a)\csc(b)$$

$$\csc(a) - \csc(b) = -2\sin\left(\frac{a}{2} - \frac{b}{2}\right)\cos\left(\frac{a}{2} + \frac{b}{2}\right)\csc(a)\csc(b).$$

Products involving the Direct Function

The product of two cosecants and the product of the cosecant and secant have the following representations:

$$\csc(a)\csc(b) = \frac{2}{\cos(a-b) - \cos(a+b)}$$

$$\csc(a) - \csc(b) = \frac{2}{\sin(a-b) + \sin(a+b)}.$$

Inequalities

Some inequalities for the cosecant function can be easily derived from the corresponding inequalities for the sine function:

$$\csc(x) > \frac{1}{x} /; x > 0.$$

$$|\csc(x)| \geq 1 /; x \in \mathbb{R}$$

$$x\csc(x) < \frac{\pi}{2} /; -\frac{\pi}{2} < x < \frac{\pi}{2}.$$

Relations with its Inverse Function

There are simple relations between the function $\csc(z)$ and its inverse function $\csc^{-1}(z)$:

$$\csc(\csc^{-1}(z)) = z \quad \csc^{-1}(\csc(z)) = z /; |\mathrm{Re}(z)| < \frac{\pi}{2} \vee \mathrm{Re}(z) = -\frac{\pi}{2} \wedge \mathrm{Im}(z) \geq 0 \vee$$

$$\mathrm{Re}(z) = \frac{\pi}{2} \wedge \mathrm{Im}(z) \geq 0.$$

The second formula is valid at least in the vertical strip $-\frac{\pi}{2} < \mathrm{Re}(z) < \frac{\pi}{2}$. Outside of this strip a much more complicated relation (that contains the unit step, real part, and the floor functions) holds:

$$\csc^{-1}(\csc(z)) = (-1)^{\left\lfloor \frac{\mathrm{Re}(z)}{2\pi} - \frac{1}{2} \right\rfloor} \left(\left(1 + (-1)^{\left\lfloor \frac{\mathrm{Re}(z)}{2\pi} + \frac{1}{2} \right\rfloor + \left\lfloor -\frac{\mathrm{Re}(z)}{\pi} - \frac{1}{2} \right\rfloor} \right) \theta(-\mathrm{Im}(z)) - 1 \right)$$

$$\left(z - \pi \left\lfloor \frac{\mathrm{Re}(z)}{\pi} + \frac{1}{2} \right\rfloor + \frac{\pi}{2} \left(1 + (-1)^{\left\lfloor \frac{\mathrm{Re}(z)}{2\pi} + \frac{1}{2} \right\rfloor + \left\lfloor \frac{\mathrm{Re}(z)}{\pi} - \frac{1}{2} \right\rfloor} \right) \theta(-\mathrm{Im}(z)) \right).$$

Representations through other Trigonometric Functions

Cosecant and secant functions are connected by a very simple formula that contains the linear function in the argument:

$$\csc(z) = \sec\left(\frac{\pi}{2} - z \right).$$

The cosecant function can also be represented using other trigonometric functions by the following formulas:

$$\csc(z)=\frac{1}{\cos\left(\frac{\pi}{2}-z\right)}\quad \cos(z)=\frac{\tan^2\left(\frac{z}{2}\right)+1}{2\tan\left(\frac{z}{2}\right)}\quad \cos(z)=\frac{\cot^2\left(\frac{z}{2}\right)+1}{2\cot\left(\frac{z}{2}\right)}.$$

Representations through Hyperbolic Functions

The cosecant function has representations using the hyperbolic functions:

$$\csc(z)=\frac{i}{\sinh(iz)}\quad \cos(z)=\frac{i}{\cosh\left(\frac{\pi i}{2}-iz\right)}\quad \csc(z)=\frac{i\left(1-\tanh^2\left(\frac{zi}{2}\right)\right)}{2\tanh\left(\frac{zi}{2}\right)}$$

$$\csc(z)=\frac{i\left(\coth^2\left(\frac{iz}{2}\right)-1\right)}{2\coth\left(\frac{iz}{2}\right)}$$

$$\csc(z)=i\operatorname{csch}(iz)\quad \cos(iz)=-i\operatorname{csch}(iz)\quad \operatorname{csch}(z)=\operatorname{sech}\left(\frac{\pi i}{2}-iz\right).$$

Applications

The cosecant function is used throughout mathematics, the exact sciences, and engineering.

Cosecant Function is Odd

Theorem:

Let $x \in \mathbb{R}$ be a real number.

Let csc x be the cosecant of x. Then,

whenever csc x is defined: csc( )

– = x x c– os

That is, the cosecant function is odd.

Proof:

$$\begin{aligned}\csc(-x) &= \frac{1}{\sin(-x)} && \text{Cosecant is Reciprocal of Sine}\\ &= \frac{1}{-\sin x} && \text{Sine Function is Odd}\\ &= -\csc x && \text{Cosecant is Reciprocal of Sine}\end{aligned}$$

Derivative of Cosecant Function

Theorem:

$$D_x(\csc x) = -\csc x \cot x$$

where $\sin x \neq 0$.

Proof:

From the definition of the cosecant function:

$$\csc x = \frac{1}{\sin x}$$

From Derivative of Sine Function:

$$D_x(\sin x) = \cos x$$

Then:

$$\begin{aligned}D_x(\sin x) &= \cos x \frac{-1}{\sin^2 x} && \text{Chain Rule}\\ &= \frac{-1}{\sin x}\frac{\cos x}{\sin x}\\ &= -\csc x \cot x && \text{Definitions of secant and cotangent}\end{aligned}$$

This is valid only when $\sin x \neq 0$.

Power Series Expansion for Cosecant Function

Theorem:

The cosecant function has a Laurent series expansion:

$$\begin{aligned}\csc x &= \sum_{n=0}^{\infty} \frac{(-1)^{n-1} 2\left(2^{2n-1}-1\right) B_{2n}\, x^{2n-1}}{(2n)!}\\ &= \frac{1}{x} + \frac{x}{6} + \frac{7x^3}{360} + \frac{31x^5}{15120} + \cdots\end{aligned}$$

where B_n denotes the Bernoulli numbers.

This converges for $0 < |x| < \pi$.

Proof:

	$\sin x = 2\sin\frac{x}{2}\cos\frac{x}{2}$	Double Angle Formula for Sine
$\Leftrightarrow$	$\frac{1}{\sin x} = \frac{1}{2\sin\frac{x}{2}\cos\frac{x}{2}}$	taking the reciprocal of both sides
$\Leftrightarrow$	$\csc x = \frac{1}{2}\csc\frac{x}{2}\sec\frac{x}{2}$	Cosecant is Reciprocal of Sine, Secant is Reciprocal of Cosine
	$= \frac{1}{2}\csc\frac{x}{2}\frac{2}{e^{i\frac{x}{2}} + e^{-i\frac{x}{2}}}$	Secant Exponential Formulation
	$= \csc\frac{x}{2}\frac{e^{-i\frac{x}{2}}}{1+e^{-ix}}$	
$\Leftrightarrow$	$\csc x\left(1+e^{-ix}\right) = \csc\frac{x}{2}e^{-i\frac{x}{2}}$	multiplying both sides by $1+e^{-ix}$
$\Leftrightarrow$	$\csc x = \csc\frac{x}{2}e^{-i\frac{x}{2}} - \csc x e^{-ix}$	subtracting $\csc x e^{-ix}$ from both sides
	$= \frac{2i}{e^{i\frac{x}{2}} - e^{-i\frac{x}{2}}}e^{-i\frac{x}{2}} - \frac{2i}{e^{ix} - e^{-ix}}e^{-ix}$	Cosecant Exponential Formulation
	$= 2\frac{i}{e^{ix}-1} - \frac{2i}{e^{2ix}-1}$	
	$= \frac{1}{x}\left(2\frac{ix}{e^{ix}-1} - \frac{2ix}{e^{2ix}-1}\right)$	
	$= \frac{1}{x}\sum_{n=0}^{\infty}\left(2\frac{(ix)^n B_n}{n!} - \frac{(2ix)^n B_n}{n!}\right)$	Definition of Bernoulli Numbers
	$= \frac{1}{x}\sum_{n=0}^{\infty}\frac{B_n\left(2(ix)^n - (2ix)^n\right)}{n!}$	
	$= \frac{1}{x}\sum_{n=0}^{\infty}\frac{B_n i^n x^{n-1} 2\left(1-2^{n-1}\right)}{n!}$	
	$= \sum_{n=0}^{\infty}\frac{B_{2n} i^n x^{2n-1} 2\left(1-2^{2n-1}\right)}{(2n)!}$	Odd Bernoulli Numbers Vanish and the term $n = 1$ vanishes
	$= \sum_{n=0}^{\infty}\frac{B_{2n}(-1)^{n-1} x^{2n-1} 2\left(2^{2n-1}-1\right)}{(2n)!}$	

Convergence

By Combination Theorem for Limits of Functions we can deduce the following:

$$\lim_{n\to\infty}\left|\frac{\dfrac{B_{2n+1}(-1)^n x^{2n+1}\left(2^{2n+1}-1\right)}{(2n+2)!}}{\dfrac{B_{2n}(-1)^{n-1}x^{2n-1}\left(2^{2n-1}-1\right)}{(2n)!}}\right| = \lim_{n\to\infty}\left|\frac{2^{2n+1}-1}{2^{2n-1}-1}\frac{1}{(2n+1)(2n+2)}\frac{B_{2n+2}}{B_{2n}}\right|x^2$$

$$= \lim_{n\to\infty}\left|\frac{2^{2n+1}-1}{2^{2n-1}-1}\right|\left|\frac{1}{(2n+1)(n+1)}\frac{B_{2n+2}}{B_{2n}}\right|\frac{1}{2}x^2$$

$$= \lim_{n\to\infty}\left|\frac{2^{2n+1}}{2^{2n-1}-1}-\frac{1}{2^{2n-1}-1}\right|\left|\frac{1}{(2n+1)(n+1)}\frac{B_{2n+2}}{B_{2n}}\right|\frac{1}{2}x^2$$

$$= \lim_{n\to\infty}\left|4\frac{2^{2n+1}-1+1}{2^{2n-1}-1}-\frac{1}{2^{2n-1}-1}\right|\left|\frac{1}{(2n+1)(n+1)}\frac{B_{2n+2}}{B_{2n}}\right|\frac{1}{2}x^2$$

$$= \lim_{n\to\infty}\left|4+\frac{4}{2^{2n-1}-1}-\frac{1}{2^{n-1}-1}\right|\left|\frac{1}{(2n+1)(n+1)}\frac{B_{2n+2}}{B_{2n}}\right|\frac{1}{2}x^2$$

$$= \lim_{n\to\infty}\left|\frac{1}{(2n+1)(n+1)}\frac{B_{2n+2}}{B_{2n}}\right|2x^2$$

$$= \lim_{n\to\infty}\left|\frac{1}{(2n+1)(n+1)}\frac{B_{2n+2}}{B_{2n}}\right|2x^2$$

$$= \lim_{n\to\infty}\left|\frac{1}{(2n+1)(n+1)}\frac{(-1)^{n+2}4\sqrt{\pi(n+1)}\left(\dfrac{n+1}{\pi e}\right)^{2n+2}}{(-1)^{n+1}4\sqrt{\pi n}\left(\dfrac{n}{\pi e}\right)^{2n}}\right|2x^2$$

$$= \lim_{n\to\infty}\left|\frac{(n+1)^2}{(2n+1)(n+1)}\sqrt{\frac{n+1}{n}}\left(\frac{n+1}{n}\right)^{2n}\right|\frac{2}{\pi^2e^2}x^2$$

$$= \lim_{n\to\infty}\left|\left(\frac{n+1}{n}\right)^{2n}\right|\frac{1}{\pi^2e^2}x^2$$

$$= \lim_{n\to\infty}\left|\left(\left(1+\frac{1}{n}\right)^n\right)^2\right|\frac{1}{\pi^2e^2}x^2$$

$$= \frac{e^2}{\pi^2e^2}x^2$$

$$= \frac{1}{\pi^2}x^2$$

This is less than 1 if $|x| < \pi$.

Hence by the Ratio Test, the outer radius of convergence is π.

The principal part of the Laurent series is finite so converges for x ≠ 0.

Thus we have the annulus of convergence to be $0 < |x| < \pi$.

Cosecant Exponential Formulation

Let z be a complex number.

Let csc z denote the cosecant function and i denote the imaginary unit $i^2 = -1$.

Then:

$$cs\,z = \frac{2i}{e^{iz} - e^{-iz}}$$

Proof:

$$\csc z = \frac{1}{\sin z} \qquad \text{Definition of Complex Cosecant Function}$$

$$= 1 / \frac{e^{iz} - e^{-iz}}{2i} \qquad \text{Sine Exponential Formulation}$$

$$= \frac{2i}{e^{iz} - e^{-iz}} \qquad \text{multiplying top and bottom by } 2i$$

Secant

The word "secant" was introduced by Alhabaš Alhãsib around 800. Later on Th. Fincke used the word "secans" in Latin for notation of the corresponding function. The secant function also appeared in the works of A. Magini and B. Cavalieri. J. Kresa used the symbol "sec" that was later widely used by L. Euler.

The classical definition of the secant function for real arguments is: "the secant of an angle $\acute{a}$ in a right-angle triangle is the ratio of the length of the hypotenuse to the adjacent leg." This description of $\sec(\acute{a})$ is valid for $0 < \acute{a} < \eth / 2$ when this triangle is non-degenerate. This approach to the secant can be expanded to arbitrary real values of $\acute{a}$ if consideration is given to the arbitrary point (x, y) in the x, y-Cartesian plane and $\sec(\acute{a})$ is defined as the ratio, assuming that α is the value of the angle between the positive direction of the x-axis and the direction from the origin to the point (x, y).

Comparing the $\sec(á)$ definition with the definition of the cosine function shows that the following formula can also be used as a definition of the secant function:

$$\sec(z) = \frac{1}{\cos(z)}.$$

Here is a graphic of the secant function $f(x) = \sec(x)$ for real values of its argument x.

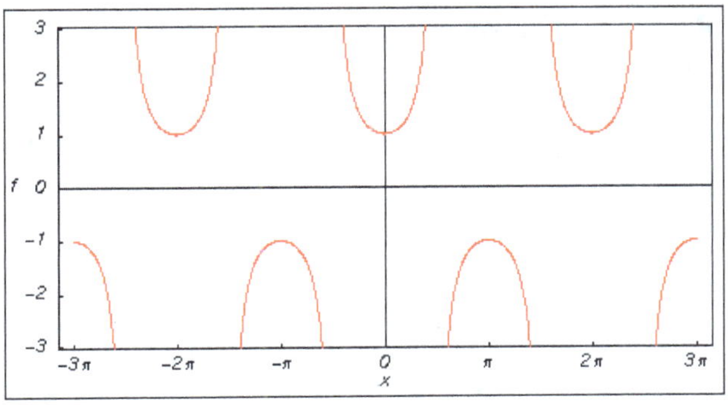

Representation through More General Functions

The secant function $\sec(z)$ can be represented using more general mathematical functions. As the ratio of one and the cosine function that is a particular case of the generalized hypergeometric, Bessel, Struve, and Mathieu functions, the secant function can also be represented as ratios of one and those special functions. Here are some examples:

$$\sec(z) = \frac{1}{{}_0F_1\left(;\frac{1}{2};-\frac{z^2}{4}\right)} \quad \sec(z) = \frac{1}{\sqrt{\frac{\eth z}{2}}\, J_{-\frac{1}{2}}(z)} \quad \sec(z) = \frac{1}{\sqrt{\frac{\eth i z}{2}}\, I_{-\frac{1}{2}}(iz)}$$

$$\sec(z) = -\frac{1}{\sqrt{\frac{\eth z}{2}}\, Y_{\frac{1}{2}}(z)} \quad \sec(z) = \frac{1}{1 - \sqrt{\frac{\eth z}{2}}\, H_{\frac{1}{2}}(z)} \quad \sec(z) = \frac{1}{Ce(1,0,z)}.$$

But these representations are not very useful because they include complicated special functions in the denominators.

It is more useful to write the secant function as particular cases of one special function. That can be done using doubly periodic Jacobi elliptic functions that degenerate into the secant function when their second parameter is equal to 0 or 1:

$$\sec(z) = \mathrm{dc}(z|0) = \mathrm{nc}(z|0) = \mathrm{ds}\left(\frac{\eth}{2} - z\middle|0\right) =$$

$$\mathrm{ns}\left(\frac{\eth}{2} - z\middle|0\right) = \mathrm{cn}(iz|1) = \mathrm{dn}(iz|1) = i\,\mathrm{cs}\left(\frac{\eth i}{2} - iz\middle|1\right) = i\,\mathrm{ds}\left(\frac{\eth i}{2} - iz\middle|1\right).$$

Definition of the Secant Function for a Complex Argument

In the complex z-plane, the function $\sec(z)$ is defined using $\cos(z)$ or the exponential function e^w in the points iz and $-iz$ through the formula:

$$\sec(z) = \frac{1}{\cos(z)} = \frac{2}{e^{iz} + e^{-iz}}.$$

In the points $\pi = \;/\pi k \;\; k\,/;\; \in \mathbb{Z}$, where $\cos(z)$ has zeros, the denominator of the last formula equals zero and $\sec(z)$ has singularities (poles of the first order).

Here are two graphics showing the real and imaginary parts of the secant function over the complex plane.

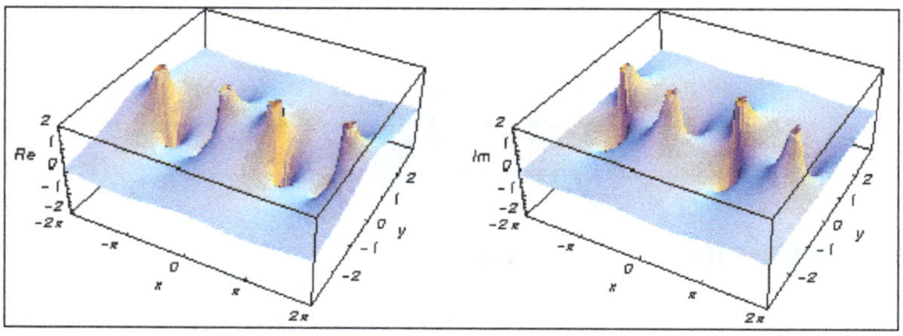

The Best-known Properties and Formulas for the Secant Function

Values in Points

Using the connection between the cosine and secant functions gives the following table of values of the secant function for angles between 0 and 2 π:

$$\sec(0) = 1 \quad \sec\left(\frac{\pi}{6}\right) = \frac{2}{\sqrt{3}} \quad \sec\left(\frac{\pi}{4}\right) = \sqrt{2} \quad \sec\left(\frac{\pi}{3}\right) = 2$$

$$\sec\left(\frac{\pi}{2}\right) = \overline{\infty} \quad \sec\left(\frac{2\pi}{3}\right) = -2 \quad \sec\left(\frac{3\pi}{4}\right) = -\sqrt{2} \quad \sec\left(\frac{5\pi}{6}\right) = -\frac{2}{\sqrt{3}}$$

$$\sec(\pi) = -1 \quad \sec\left(\frac{7\pi}{6}\right) = -\frac{2}{\sqrt{3}} \quad \sec\left(\frac{5\pi}{4}\right) = -\sqrt{2} \quad \sec\left(\frac{4\pi}{3}\right) = -2$$

$$\sec\left(\frac{3\pi}{2}\right) = \overline{\infty} \quad \sec\left(\frac{5\pi}{3}\right) = 2 \quad \sec\left(\frac{7\pi}{4}\right) = \sqrt{2} \quad \sec\left(\frac{11\pi}{6}\right) = \frac{2}{\sqrt{3}}$$

$$\sec(2\pi) = 1 \quad \sec(\pi m) = (-1)^m \,/; m \in \mathbb{Z} \quad \sec\left(\pi\left(\frac{1}{2} + m\right)\right) = \overline{\infty} \,/; m \in \mathbb{Z}.$$

General Characterists

For real values of argument z, the values of $\sec(z)$ are real.

In the points $z = 2\pi n / m /; n \in \mathbb{Z} \wedge m \in \mathbb{Z}$, $\sec(z)$ are algebraic. In several cases they can be integers -2, -1, 1 or 2:

$$\sec(0)=1 \quad \sec\left(\frac{\pi}{3}\right)=2 \quad \sec\left(\frac{2\pi}{3}\right)=-2 \quad \sec(\pi)=-1.$$

The values of $\sec\left(\frac{n\pi}{m}\right)$ can be expressed using only square roots if $n \in \mathbb{Z}$ and m is a product of a power of 2 and distinct Fermat primes {3, 5, 17, 257, ...}.

The function $\sec(z)$ is an analytical function of z that is defined over the whole complex z-plane and does not have branch cuts and branch points. It has an infinite set of singular points:

- $z = \pi/2 + \pi k /; k \in \mathbb{Z}$, are the simple poles with residues $(-1)^{k-1}$.
- $z = \tilde{\infty}$ is an essential singular point.

It is a periodic function with a real period 2π.

$$\sec(z+2\pi) = \sec(z)$$
$$\sec(z) = \sec(z+2\pi k) /; k \in \mathbb{Z}$$
$$\sec(z) = (-1)^k \sec(z+\pi k) /; k \in \mathbb{Z}.$$

The function $\sec(z)$ is an even function with mirror symmetry:

$$\sec(-z) = \sec(z) \quad \sec(\bar{z}) = \overline{\sec(z)}.$$

Differentiation

The first derivative of $\sec(z)$ has simple representations using either the $\tan(z)$ function or the $\sec(z)$ function:

$$\frac{\partial \sec(z)}{\partial z} = \tan(z)\sec(z).$$

The nth derivative of $\sec(z)$ has much more complicated representations than symbolic n^{th} derivatives for $\sec(z)$ and $\cos(z)$:

$$\frac{\partial \sec(z)}{\partial z} = \sec(z)\left(\delta_n + (n+1)! \sum_{k=0}^{n} \sum_{j=0}^{\left\lceil \frac{k-1}{2} \right\rceil} \frac{(-1)^k 2^{1-k} (k-2j)^n \sec^k(z)}{(k+1) j! (k-j)! (n-k)!} \cos\left(\frac{\pi n}{2} + (k-2j)z\right)\right) /; n \in \mathbb{N}.$$

where δ_n is the Kronecker delta symbol: $\delta_0 = 1$ and $\delta_n = 0 /; n \neq 0$.

Ordinary Differential Equation

The function $\sec(z)$ satisfies the following first-order nonlinear differential equation:

$$w'(z)^2 - w(z)^4 + w(z)^2 = 0 /; w(z) = \sec(z).$$

Series Representation

The function $\sec(z)$ has the following series expansion at the origin that converges for all finite values z with $|z| < \frac{\pi}{2}$:

$$\sec(z) = 1 + \frac{z^2}{z} + \frac{5z^4}{24} + \ldots = \sum_{k=0}^{\infty} \frac{(-1)^k E_{2k} z^{2k}}{(2k)!},$$

where E_{2k} are the Euler numbers.

The secant function $\sec(z)$ can also be presented using other kinds of series by the following formulas:

$$\sec(z) = \pi \sum_{k=0}^{\infty} \frac{(-1)^k (2k+1)}{\left(k + \frac{1}{2}\right)^2 \pi^2 - z^2} /; \frac{z}{\pi} - \frac{1}{2} \notin \mathbb{Z}$$

$$\sec^2(z) = \pi \sum_{k=-\infty}^{\infty} \frac{1}{\left(z + \pi\left(k + \frac{1}{2}\right)\right)^2} /; \frac{z}{\pi} - \frac{1}{2} \notin \mathbb{Z}.$$

Integral Representation

The function $\sec(z)$ has a well-known integral representation through the following definite integral along the positive part of the real axis:

$$\sec(z) = \frac{2}{\pi} \int_0^{\infty} \frac{t^{\frac{2z}{\pi}}}{t^2 + 1} dt /; |\mathrm{Re}(z)| < \frac{\pi}{2}.$$

Product Representation

The famous infinite product representation for $\cos(z)$ can be easily rewritten as the following product representation for the secant function:

$$\sec(z) = \prod_{k=1}^{\infty} \frac{\pi^2 (2k-1)^2}{\pi^2 (2k-1)^2 - 4z^2}.$$

Limit Representation

The secant function has the following limit representation:

$$\sec(z) = \lim_{n\to\infty} \sum_{k=-n}^{\infty} \frac{(-1)^k}{\pi\left(k+\frac{1}{2}\right)-z} /; \frac{z}{\pi}-\frac{1}{2} \notin \mathbb{Z}.$$

Indefinite Integration

Indefinite integrals of expressions involving the secant function can sometimes be expressed using elementary functions. However, special functions are frequently needed to express the results even when the integrands have a simple form (if they can be evaluated in closed form). Here are some examples:

$$\int \sec(z)\,dz = \log\left(\cot\left(\frac{z}{2}-\frac{\pi}{4}\right)\right)$$

$$\int \sqrt{\sec(z)}\,dz = 2\cos^{\frac{1}{2}}(z)\,F\left(\frac{z}{2}\middle|2\right)\sec^{\frac{1}{2}}(z)$$

$$\int \sec^{\nu}(a\,z)\,dz = \frac{\csc(a\,z)\sec^{\nu}(a\,z)\sqrt{\sin^2(a\,z)}}{a\nu-a}\,{}_2F_1\left(\frac{1-\nu}{2},\frac{1}{2};\frac{3-\nu}{2};\cos^2(a\,z)\right).$$

Definite Integration

Definite integrals that contain the secant function are sometimes simple and their values can be expressed through elementary functions. Here is one example:

$$\int_0^{\frac{\pi}{4}} t\sec^2(t)\,dt = \frac{1}{4}(\pi-\log(4)).$$

Some special functions can be used to evaluate more complicated definite integrals. For example, polygamma and gamma functions and the Catalan constant are needed to express the following integrals:

$$\int_0^{\frac{\pi}{4}} t\sec(t)\,dt = \frac{1}{32}\left(8\pi\left(\log\left(1-(-1)^{3/4}\right)-\log\left(1+(-1)^{3/4}\right)\right)+\sqrt[4]{-1}\left(\psi^{(1)}\left(\frac{1}{8}\right)-i\psi^{(1)}\left(\frac{3}{8}\right)\right.\right.$$
$$\left.\left.-\psi^{(1)}\left(\frac{5}{8}\right)+i\psi^{(1)}\left(\frac{7}{8}\right)-64c\right)\right)$$

$$\int_0^{\frac{\pi}{4}} \sec^{a}(t)\,dt = \frac{\sqrt{\pi}}{2\Gamma\left(1-\frac{a}{2}\right)}\Gamma\left(\frac{1-a}{2}\right) /; \mathrm{Re}(a)<1.$$

Finite Summation

Finite sums that contain the secant function have the following simple values:

$$\sum_{k=0}^{n} \sec\left(\frac{2k\pi}{2n+1}\right) = \frac{1}{2}(-1)^n (2n+1) + \frac{1}{2} /; n \in \mathbb{N}$$

$$\sum_{k=1}^{\left\lfloor \frac{n-1}{2} \right\rfloor} \sec^2\left(\frac{k\pi}{n}\right) = \frac{1}{12}\left(4n^2 - (-1)^n (2n^2+1) - 7\right) /; n \in \mathbb{N}^+$$

$$\sum_{k=0}^{n-1} \sec^2\left(\frac{\pi k}{n} + z\right) = n^2 \csc^2\left(\frac{1}{2} n (2z+\pi)\right) /; n \in \mathbb{N}^+$$

$$\sum_{k=0}^{n-1} \sec^2\left(\frac{2\pi k}{n} + z\right) = \frac{1}{2}\left(1-(-1)^n\right) \mathrm{n}^2 \csc^2\left(\frac{1}{2} n (2z+\pi)\right) + \frac{1}{4}\left(1+(-1)^n\right) \mathrm{n}^2$$

$$\csc^2\left(\frac{1}{4} n (2z+\pi)\right) /; n \in \mathbb{N}^+$$

$$\sum_{k=1}^{n} \frac{1}{2^{2k}} \sec^2\left(\frac{z}{2^k}\right) = \csc^2(z) - \frac{1}{2^{2n}} \csc^2\left(\frac{z}{2^n}\right) /; n \in \mathbb{N}.$$

Infinite Summation

The evaluation limit of the last formula in the previous subsubsection for $n -> \infty$ gives the following value for the corresponding infinite sum:

$$\sum_{k=1}^{\infty} \frac{1}{2^{2k}} \sec^2\left(\frac{z}{2^k}\right) = \csc^2(z) - z\frac{1}{2^2}.$$

Finite Products

The following finite product from the secant can be represented through the cosecant function:

$$\prod_{k=0}^{n-1} \sec\left(z + \frac{\pi k}{n}\right) = (-1)^{n-1} 2^{n-1} \csc\left(\frac{n(\pi - 2z)}{2}\right) /; \mathrm{n} \in \mathbb{N}^+.$$

Infinite Products

The following infinite product from the secant can be represented through the cosecant function:

$$\prod_{k=1}^{\infty} \sec\left(\frac{z}{2^k}\right) = z \csc(z).$$

Addition Formulas

The secants of a sum and a difference can be represented by the following formulas that are derived from the cosines of a sum and a difference:

$$\sec(a+b) = \frac{1}{\cos(b)\cos(a) - \sin(a)\sin(b)}$$

$$\sec(a-b) = \frac{1}{\cos(a)\cos(b) + \sin(a)\sin(b)}.$$

Multiple Arguments

In the case of multiple arguments z, 2z, 3z, ..., the function $\sec(z)$ can be represented as a rational function including powers of a secant. Here are two examples:

$$\sec(2z) = \frac{\sec^2(z)}{2 - \sec^2(z)}$$

$$\sec(3z) = \frac{\sec^3(z)}{4 - 3\sec^2(z)}.$$

Half-angle Formulas

The secant of a half-angle can be represented by the following simple formula that is valid in a vertical strip:

$$\sec\left(\frac{z}{2}\right) = \frac{\sqrt{2}}{\sqrt{1+\cos(z)}} /; |\mathrm{Re}(z)| < \pi \vee \mathrm{Re}(z) = -\pi \wedge \mathrm{Im}(z) > 0 \vee \mathrm{Re}(z) = \pi \wedge \mathrm{Im}(z) < 0.$$

To make this formula correct for all complex z, a complicated prefactor is needed:

$$\sec\left(\frac{z}{2}\right) = \frac{c(z)}{\sqrt{1+\cos(z)}} /; c(z) = (-1)^{\left\lfloor \frac{\mathrm{Re}(z)+\pi}{2\pi} \right\rfloor} \sqrt{2}\left(1 - \left(1 + (-1)^{\left\lfloor \frac{\mathrm{Re}(z)+\pi}{2\pi} \right\rfloor + \left\lfloor -\frac{\mathrm{Re}(z)+\pi}{2\pi} \right\rfloor}\right)\theta(-\mathrm{Im}(z))\right),$$

where $c(z)$ contains the unit step, real part, imaginary part, and the floor functions.

Sums of Two Direct Functions

The sum and difference of two secant functions can be described by the following formulas:

$$\sec(a) + \sec(b) = 2\cos\left(\frac{a}{2} - \frac{b}{2}\right)\cos\left(\frac{a}{2} + \frac{b}{2}\right)\sec(a)\sec(b)$$

$$\sec(a) - \sec(b) = 2\sin\left(\frac{a}{2} - \frac{b}{2}\right)\sin\left(\frac{a}{2} + \frac{b}{2}\right)\sec(a)\sec(b).$$

Products Involving the Direct Function

The product of two secants and the product of a secant and a cosecant have the following representations:

$$\sec(a)\sec(b) = \frac{1}{\cos(a-b)+\cos(a+b)}$$

$$\sec(a)\csc(b) = \frac{1}{\sin(a+b)-\sin(a-b)}.$$

Inequalities

One of the most famous inequalities for a secant function is the following:

$$\sec(x) > x\csc(x)/;0<x<\frac{\pi}{2}\wedge x\in\mathbb{R}.$$

Relations with its Inverse Function

There are simple relations between the function $\sec(z)$ and its inverse function $\sec^{-1}(z)$:

$$\sec\left(\sec^{-1}(z)\right)=z \quad \sec^{-1}\left(\sec(z)\right)=z/;0<\mathrm{Re}(z)<\pi\vee\mathrm{Re}(z)=0\wedge\mathrm{Im}(z)\geq 0\vee\mathrm{Re}(z)$$
$$=\pi\wedge\mathrm{Im}(z)\leq 0.$$

The second formula is valid at least in the vertical strip $0<\mathrm{Re}(z)<\pi$. Outside of this strip a much more complicated relation (containing the unit step, real part, and the floor functions) holds:

$$\sec^{-1}\left(\sec(z)\right)=\frac{\pi}{2}\left(1-(-1)^{\left\lfloor -\frac{\mathrm{Re}(z)}{\pi}\right\rfloor}\right)+(-1)^{\left\lfloor -\frac{\mathrm{Re}(z)}{\pi}\right\rfloor}\left(\left(1+(-1)^{\left\lfloor \frac{\mathrm{Re}(z)}{\pi}\right\rfloor+\left\lfloor -\frac{\mathrm{Re}(z)}{\pi}\right\rfloor}\right)\theta\left(\mathrm{Im}(z)\right)-1\right)$$
$$\left(z+\pi\left\lfloor -\frac{\mathrm{Re}(z)}{\pi}\right\rfloor\right).$$

Representations through other Trigonometric Functions

Secant and cosecant functions are connected by a very simple formula that contains the linear function in the argument:

$$\sec(z)=\csc\left(\frac{\pi}{2}-z\right).$$

The secant function can also be represented using other trigonometric functions by the following formulas:

$$\sec(z)=\frac{1}{\sin\left(\frac{\pi}{2}-z\right)}\quad \sec(z)=\frac{1+\tan^2\left(\frac{z}{2}\right)}{1-\tan^2\left(\frac{z}{2}\right)}\quad \sec(z)=\frac{\cot^2\left(\frac{z}{2}\right)+1}{\cot^2\left(\frac{z}{2}\right)-1}.$$

Representations through Hyperbolic Functions

The secant function has representations using the hyperbolic functions:

$$\sec(z)=\frac{i}{\sin\left(\frac{\pi i}{2}-iz\right)}\quad \sec(z)=\frac{1}{\cosh(iz)}\quad \sec(z)=\frac{1-\tan^2\left(\frac{iz}{2}\right)}{1+\tan^2\left(\frac{iz}{2}\right)}\quad \sec(z)=\frac{\coth^2\left(\frac{iz}{2}\right)-1}{\coth^2\left(\frac{iz}{2}\right)+1}$$

$$\sec(z)=i\operatorname{csch}\left(\frac{\pi i}{2}-iz\right)\quad \sec(z)=\operatorname{sech}(iz)\quad \sec(iz)=\sec h(z).$$

Applications

The secant function is used throughout mathematics, the exact sciences, and engineering.

Secant Function is Even

Theorem:

Let $x\in\mathbb{R}$ be a real numberLet

sec x be the secant of x. Then,

whenever sec x is defined:

$$\sec(-x)=\sec x$$

That is, the secant function is even.

Proof:

$$\sec(-x)=\frac{1}{\cos(-x)}\quad \text{Secant is Reciprocal of Cosine}$$

$$=\frac{1}{\cos x}\quad \text{Cosine Function is Even}$$

$$=\sec x\quad \text{Secant is Reciprocal of Cosine}$$

Derivative of Secant Function

Theorem:

$$D_x(\sec x) = \sec x \tan x$$

where $\cos x \neq 0$.

Proof:

From the definition of the secant function:

$$\sec x = \frac{1}{\cos x} = (\cos x)^{-1}$$

From Derivative of Cosine Function:

$$D_x(\cos x) = -\sin x$$

Then:

$D_x(\sec x) = D_x\left((\cos x)^{-1}\right)$	Exponent Laws
$= (-\sin x)(-\cos^{-2} x)$	Chain Rule, Power Rule
$= \frac{1}{\cos x}\frac{\sin x}{\cos x}$	Exponent Laws
$= \sec x \tan x$	Definitions of secant and tangent

This is valid only when $\cos x \neq 0$.

Power Series Expansion for Secant Function

Theorem

The (real) secant function has a Taylor series expansion:

$$\sec x = \sum_{n=1}^{\infty}(-1)\frac{E_{2n}\, x^{2n}}{(2n)!}$$
$$= 1 + \frac{x^2}{2} + \frac{5x^4}{24} + \frac{61x^6}{720} + \cdots$$

where E_{2n} denotes the Euler numbers.

This converges for $|x| < \frac{\pi}{2}$.

Secant Exponential Formulation

Theorem:

Let z be a complex number.

Let sec z denote the secant function and i denote the imaginary unit: $i^2 = -1$.

Then:

$$\sec z = \frac{2}{e^{iz} + e^{-iz}}$$

Proof:

$$\sec z = \frac{1}{\cos z}$$ Definition of Complex Secant Function

$$= 1 / \frac{e^{iz} + e^{-iz}}{2i}$$ Sine Exponential Formulation and Cosine Exponential Formulation

$$= \frac{2}{e^{iz} + e^{-iz}}$$ multiplying top and bottom by 2

Cotangent

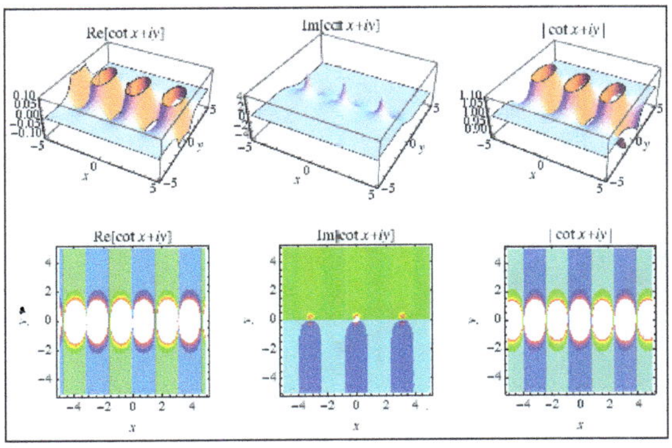

The cotangent function cot z is the function defined by,

$$\cot z = \frac{1}{\tan z}$$

$$= \frac{i\left(e^{iz} + e^{-iz}\right)}{e^{iz} - e^{-iz}}$$

$$= \frac{i\left(e^{2iz} + 1\right)}{e^{2iz} - 1},$$

where tan z is the tangent.

The notations ctn z and ctg z are sometimes used in place of cotz. Note that the cotangent is not in as widespread use in Europe as are sin z, cos z, and tan z, although it does appear explicitly in various German and Russian handbooks. Interestingly, cotz is treated on par with the other trigonometric functions in most tabulations while sec z and csc z are sometimes not.

An important identity connecting the cotangent with the cosecant is given by,

$$1+\cot^2 \grave{e} = \cot^2 \grave{e}.$$

The cotangent has smallest real fixed point x such cot x = x at 0.8603335890....

The derivative is given by,

$$\frac{d}{dz}\cot z = -\cot^2 z$$

and the indefinite integral by,

$$\int \cot z \, d\, z = \ln(\sin z) + C,$$

where C is a constant of integration.

Definite integrals include,

$$\int_{\pi/4}^{\pi/2} \cot x \, d\, x = \frac{1}{2}\ln 2$$

$$\int_0^{\pi/4} \ln(\cot x) \, d\, x = K$$

$$\int_0^{\pi/4} x\cot x \, d\, x = \frac{1}{8}(\pi \ln 2 + 4K)$$

$$\int_0^{\pi/2} x\cot x \, d\, x = \frac{1}{2}\pi \ln 2$$

$$\int_{\pi/4}^{\pi/2} x\cot x \, d\, x = \frac{1}{8}(3\pi \ln 2 - 4K)$$

$$\int_0^{\pi/4} x^2 \cot x \, d\, x = \frac{1}{64}\left[16\pi K + 2\pi^2 \ln 2 - 34\zeta(3)\right]$$

$$\int_0^{\pi/2} x^2 \cot x \, d\, x = \frac{1}{8}\left[2\pi^2 \ln 2 - 7\zeta(3)\right],$$

where K is Catalan's constant, ln 2 is the natural logarithm of 2, and ζ (3) is Apéry's constant. Integrals $\int_0^{\pi/4} x\cot x \, d\, x = \frac{1}{8}(\pi \ln 2 + 4K)$ and $\int_0^{\pi/2} x\cot x \, d\, x = \frac{1}{2}\pi \ln 2$ were considered by Glaisher. Additional integrals include,

$$\int_0^{\pi/4} \cot^n x \, d\, x = \frac{1}{4}\left[\psi_0\left(\frac{1}{4}(3-n)\right) - \psi_0\left(\frac{1}{4}(1-n)\right)\right]$$

for R[n]<1, where psi $\phi_0(z)$ is the digamma function, and

$$\int_0^{\pi/2} \cot^n x\,dx = \frac{1}{2}\pi \sec\left[\frac{1}{2}(\pi n)\right]$$

for $-1 < R[n] < 1$.

The Laurent series for cot z about the origin is

$$\cot z = \frac{1}{z} - \frac{1}{3}z - \frac{1}{45}z^3 - \frac{2}{945}z^5 - \frac{1}{4725}z^7 - \dots$$
$$= \sum_{n=0}^{\infty} \frac{(-1)^n 2^{2n} B_{2n}}{(2n)!} z^{2n-1}$$

where B_n is a Bernoulli number.

A nice sum identity for the cotangent is given by,

$$\pi \cot(\pi z) = \frac{1}{z} + 2z \sum_{n=1}^{\infty} \frac{1}{z^2 - n^2}.$$

For an integer $n \geq 3$, cot (π / n) is rational only for n = 4. In particular, the algebraic degrees of cot (π / n) for n=2, 3, ... are 1, 2, 1, 4, 2, 6, 2, 6, 4, 10, 2, ...

Cotangent Function is Odd

Theorem:

Let $x \in \mathbb{R}$ be a real number.

Let cot x be the cotangent of x.

Then, whenever cot x is defined:

$$\cot(-x) = -\cot x$$

That is, the cotangent function is odd.

Proof:

$\cot(-x) = \frac{\cos(-x)}{\sin(-x)}$	Cotangent is Cosine divided by Sine
$= \frac{-\sin x}{\cos x}$	Cosine Function is Even and Sine Function is Odd
$= -\cot x$	Cotangent is Cosine divided by Sine

Derivative of Cotangent Function

Theorem:

$$D_x(\cot x) = -\csc^2 x = \frac{-1}{\sin^2 x}$$

where $\sin x \neq 0$.

Corollary:

$$D_x(\cot ax) = -a\csc^2 ax$$

Proof:

From the definition of the cotangent function:

$$\cot x = \frac{\cos x}{\sin x}$$

From Derivative of Sine Function:

$$D_x(\sin x) = \cos x$$

From Derivative of Cosine Function:

$$D_x(\cos x) = -\sin x$$

Then:

$$D_x(\cot x) = \frac{\sin x(-\sin x) - \cos x \cos x}{\sin^2 x} \quad \text{Quotient Rule for Derivatives}$$

$$= \frac{-\sin^2 x + \cos^2 x}{\sin^2 x}$$

$$= \frac{-1}{\sin^2 x} \quad \text{Sum of Squares of Sine and Cosine}$$

This is valid only when $\sin x \neq 0$.

The result follows from the definition of the cosecant function.

Power Series Expansion for Cotangent Function

Theorem:

The (real) cotangent function has a Taylor series expansion:

$$\cot x = \sum_{n=0}^{\infty} \frac{(-1)^n 2^{2n} B_{2n} x^{2n-1}}{(2n)!}$$
$$= \frac{1}{x} - \frac{x}{3} - \frac{x^3}{45} - \frac{2x^5}{945} + \cdots$$

where B_{2n} denotes the Bernoulli numbers.

This converges for $0 < |x| < \pi$.

Proof:

$$\cot x = i\frac{e^{ix} + e^{-ix}}{e^{ix} - e^{-ix}} \qquad \text{Cotangent Exponential Formulation}$$
$$= i\frac{e^{2ix} + 1}{e^{2ix} - 1}$$
$$= i\left(1 + \frac{2}{e^{2ix} - 1}\right)$$
$$= i + \frac{2i}{e^{2ix} - 1}$$
$$= i + \frac{1}{x}\frac{2ix}{e^{2ix} - 1}$$
$$= i + \frac{1}{x}\sum_{n=0}^{\infty} \frac{B_n (2ix)^n}{n!} \qquad \text{Definition of Bernoulli Numbers}$$
$$= \frac{1}{x} + \frac{1}{x}\sum_{n=2}^{\infty} \frac{B_n (2ix)^n}{n!} \qquad \text{as } B_0 = 1 \text{ and } B_1 = -\frac{1}{2}$$
$$= \frac{1}{x} + \frac{1}{x}\sum_{n=1}^{\infty} \frac{B_n (2ix)^{2n}}{(2n)!} \qquad \text{Odd Bernoulli Numbers Vanish}$$
$$= \frac{1}{x}\sum_{n=0}^{\infty} \frac{B_n (2ix)^{2n}}{(2n)!}$$
$$= \sum_{n=0}^{\infty} \frac{(-1)^n 2^{2n} B_{2n} x^{2n-1}}{(2n)!}$$

By Combination Theorem for Limits of Functions we can deduce the following:

$$\lim_{n\to\infty}\left|\frac{\dfrac{(-1)^n 2^{n+2} B_{2n+2}\, x^{2n+1}}{(2n+2)!}}{\dfrac{(-1)^{n-1} 2^{2n} B_{2n}\, x^{2n-1}}{(2n)!}}\right|$$

$$= \lim_{n\to\infty}\left|\frac{1}{(2n+1)(2n+2)}\frac{B_{2n+2}}{B_{2n}}\right| 4x^2$$

$$= \lim_{n\to\infty}\left|\frac{1}{(2n+1)(n+1)}\frac{B_{2n+2}}{B_{2n}}\right| 2x^2$$

$$= \lim_{n\to\infty}\left|\frac{1}{(2n+1)(n+1)}\frac{(-1)^{n+2}\sqrt[4]{\pi(n+1)}\left(\dfrac{n+1}{\pi e}\right)^{2n+2}}{(-1)^{n+1}\sqrt[4]{\pi n}\left(\dfrac{n}{\pi e}\right)^{2n}}\right| 2x^2$$ Asymptotic Formula for Bernoulli Numbers

$$=$$

$$\lim_{n\to\infty}\left|\frac{(n+1)^2}{(2n+1)(n+1)}\sqrt{\frac{n+1}{n}}\left(\frac{n+1}{n}\right)^{2n}\right|\frac{2}{\pi^2 e^2}x^2$$

$$= \lim_{n\to\infty}\left|\left(\frac{n+1}{n}\right)^{2n}\right|\frac{x^2}{\pi^2 e^2}$$

$$= \lim_{n\to\infty}\left|\left(\left(1+\frac{1}{n}\right)^n\right)^2\right|\frac{x^2}{\pi^2 e^2}$$

$$= \frac{e^2 x^2}{\pi^2 e^2}$$ Definition of Euler's Number

$$= \frac{x^2}{\pi^2}$$

This is less than 1 if and only if:

$$|x| < \pi.$$

Hence by the Ratio Test, the series converges for $|x| < \pi$.

Cotangent Exponential Formulation

Theorem:

Let z be a complex number.

Let cot z denote the cotangent function and i denote the imaginary unit: $i^2 = -1$.

Then:

$$\cot z = i\frac{e^{iz} + e^{-iz}}{e^{iz} - e^{-iz}}$$

Proof:

$$\cot z = \frac{\cos z}{\sin z} \qquad \text{Definition of Complex Contagent Function}$$

$$= \frac{e^{iz} + e^{-iz}}{2} / \frac{e^{iz} - e^{-iz}}{2i} \qquad \text{Sine Exponential Formulation and Cosine Exponential Formulation}$$

$$= i\frac{e^{iz} + e^{-iz}}{e^{iz} - e^{-iz}} \qquad \text{multiplying numerator and denominator by } 2i$$

Proof:

$$\cot z = \frac{1}{\tan z} \qquad \text{Definition of Complex Cotangent Function}$$

$$= 1 / \frac{e^{iz} - e^{-iz}}{i\left(e^{iz} + e^{-iz}\right)} \qquad \text{Tangent Exponential Formulation/ Formulation 2}$$

$$= i\frac{e^{iz} + e^{-iz}}{e^{iz} - e^{-iz}}$$

Also defined as:

This result is sometimes also presented as:

$$\cot z = \frac{i\left(e^{iz} + e^{-iz}\right)}{e^{iz} - e^{-iz}}$$

Trigonometric Functions and the Unit Circle

A unit circle is a circle centered at the origin with radius 1. The angle t (in radians) forms an arc of length s.

The x- and y-axes divide the coordinate plane (and the unit circle, since it is centered at the origin) into four quarters called quadrants. We label these quadrants to mimic the direction a positive angle would sweep. The four quadrants are labeled I, II, III, and IV.

For any angle t, we can label the intersection of its side and the unit circle by its coordinates, (x, y). The coordinates x and y will be the outputs of the trigonometric functions $f(t) = \cos t$ and $f(t) = \sin t$, respectively. This means:

$$x = \cos t$$

$$y = \sin t$$

The diagram of the unit circle illustrates these coordinates.

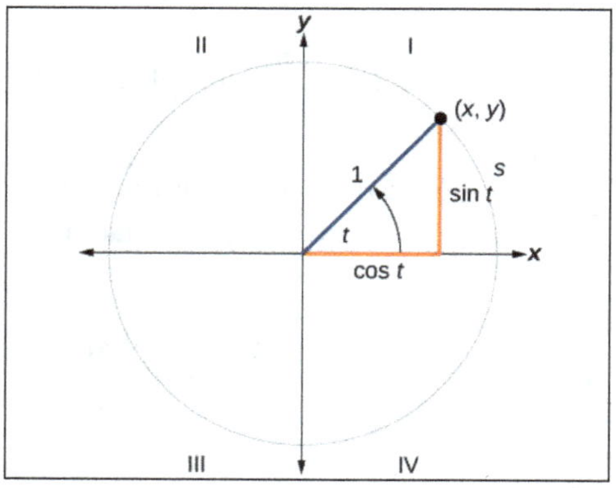

Unit circle: Coordinates of a point on a unit circle where the central angle is t radians.

The values of x and y are given by the lengths of the two triangle legs that are colored red. This is a right triangle, and you can see how the lengths of these two sides (and the values of x and y) are given by trigonometric functions of t.

For an example of how this applies, consider the diagram showing the point with coordinates $-\frac{\sqrt{2}}{2}, \frac{\sqrt{2}}{2}$ on a unit circle.

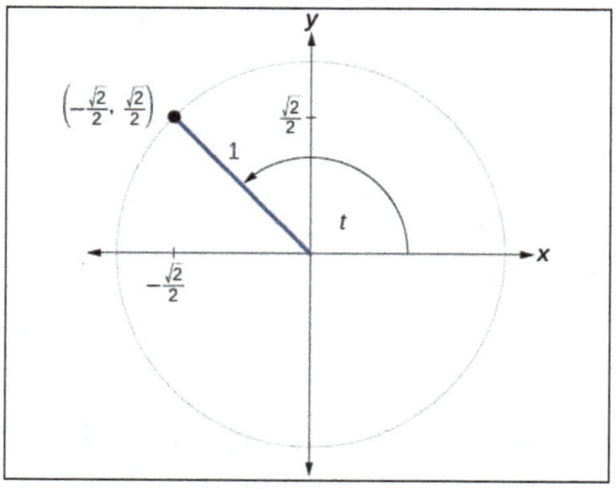

Point on a unit circle: The point $-\frac{\sqrt{2}}{2}, \frac{\sqrt{2}}{2}$ on a unit circle.

We know that, for any point on a unit circle, the x-coordinate is cos t and the y-coordinate is sin t. Applying this, we can identify that $t = -\frac{\sqrt{2}}{2}$ and $\sin t = -\frac{\sqrt{2}}{2}$ for the angle t in the diagram.

Recall that tan $t = \frac{\sin t}{\cos t}$. Applying this formula, we can find the tangent of any angle identified by a unit circle as well. For the angle t identified in the diagram of the unit circle showing the point $\left(-\frac{\sqrt{2}}{2}, \frac{\sqrt{2}}{2}\right)$, the tangent is:

$$\tan t = \frac{\sin t}{\cos t}$$
$$= \frac{-\frac{\sqrt{2}}{2}}{-\frac{\sqrt{2}}{2}}$$
$$= 1$$

We have previously discussed trigonometric functions as they apply to right triangles. This allowed us to make observations about the angles and sides of right triangles, but these observations were limited to angles with measures less than 90°. Using the unit circle, we are able to apply trigonometric functions to angles greater than 90°.

Further Consideration of the Unit Circle

The coordinates of certain points on the unit circle and the measure of each angle in radians and degrees are shown in the unit circle coordinates diagram. This diagram allows one to make observations about each of these angles using trigonometric functions.

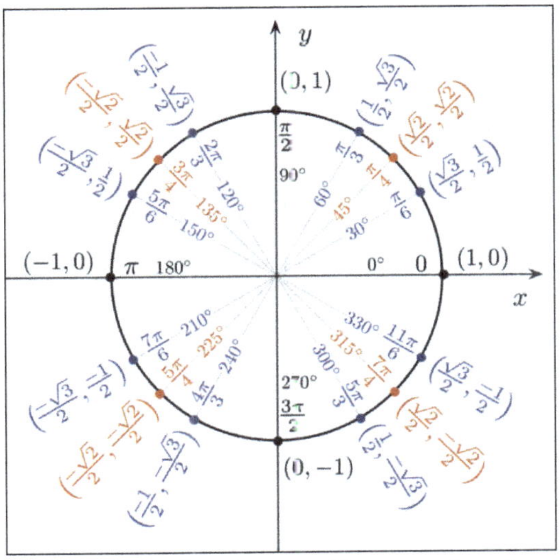

Unit circle coordinates: The unit circle, showing coordinates and angle measures of certain points.

We can find the coordinates of any point on the unit circle. Given any angle t, we can find the x– or y-coordinate at that point using x = cos t and y = sin t.

The unit circle demonstrates the periodicity of trigonometric functions. Periodicity refers to the way trigonometric functions result in a repeated set of values at regular intervals. Take a look at the x-values of the coordinates in the unit circle above for values of t from 0 to 2π:

$$1,\frac{\sqrt{3}}{2},\frac{\sqrt{2}}{2},\frac{1}{2},0,-\frac{1}{2},-\frac{\sqrt{2}}{2},-\frac{\sqrt{3}}{2},-1,-\frac{\sqrt{3}}{2},-\frac{\sqrt{2}}{2},-\frac{1}{2},0,\frac{1}{2},\frac{\sqrt{2}}{2},\frac{\sqrt{3}}{2},1$$

We can identify a pattern in these numbers, which fluctuate between −1 and 1. Note that this pattern will repeat for higher values of t. Recall that these x-values correspond to cos t. This is an indication of the periodicity of the cosine function.

Example:

$$\text{Solve in}\left(\frac{7\pi}{6}\right).$$

It seems like this would be complicated to work out. However, notice that the unit circle diagram shows the coordinates at $t=\frac{7\pi}{6}$. Since the y-coordinate corresponds to sin t, we can identify that,

$$\sin=\frac{7\pi}{6}=-\frac{1}{2}$$

Special Angles

The unit circle and a set of rules can be used to recall the values of trigonometric functions of special angles.

Trigonometric Functions of Special Angles

Recall that certain angles and their coordinates, which correspond to x = cos t and y = sin t for a given angle t, can be identified on the unit circle.

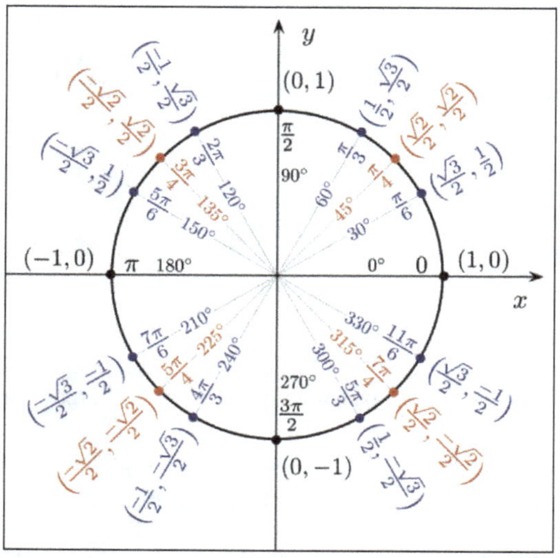

Unit circle: Special angles and their coordinates are identified on the unit circle.

The angles identified on the unit circle above are called special angles; multiples of $\pi, \frac{\pi}{2}, \frac{\pi}{3}, \frac{\pi}{4}$, and (180°, 90°, 60°, 45°, and 30°). These have relatively simple expressions. Such simple expressions generally do not exist for other angles. Some examples of the algebraic expressions for the sines of special angles are:

$$\sin(0°) = 0$$

$$\sin(30°) = \frac{1}{2}$$

$$\sin(45°) = \frac{\sqrt{2}}{2}$$

$$\sin(60°) = \frac{\sqrt{3}}{2}$$

$$\sin(90°) = 1$$

The expressions for the cosine functions of these special angles are also simple.

Note that while only sine and cosine are defined directly by the unit circle, tangent can be defined as a quotient involving these two:

$$\tan t = \frac{\sin t}{\cos t}.$$

Tangent functions also have simple expressions for each of the special angles.

We can observe this trend through an example. Let's find the tangent of 60°.

First, we can identify from the unit circle that:

$$\sin(60°) = \frac{\sqrt{3}}{2}$$

$$\cos(60°) = \frac{1}{2}$$

We can easily calculate the tangent:

$$\begin{aligned}\tan(60°) &= \frac{\sin(60°)}{\cos(60°)} \\ &= \frac{\frac{\sqrt{3}}{2}}{\frac{1}{2}} \\ &= \frac{\sqrt{3}}{2} \cdot \frac{2}{1} \\ &= \sqrt{3}\end{aligned}$$

Memorizing Trigonometric Functions

An understanding of the unit circle and the ability to quickly solve trigonometric functions for certain angles is very useful in the field of mathematics. Applying rules and shortcuts associated with the unit circle allows you to solve trigonometric functions quickly. The following are some rules to help you quickly solve such problems.

Signs of Trigonometric Functions

The sign of a trigonometric function depends on the quadrant that the angle falls in. To help remember which of the trigonometric functions are positive in each quadrant, we can use the mnemonic phrase "A Smart Trig Class." Each of the four words in the phrase corresponds to one of the four quadrants, starting with quadrant I and rotating counterclockwise. In quadrant I, which is "A," all of the trigonometric functions are positive. In quadrant II, "Smart," only sine is positive. In quadrant III, "Trig," only tangent is positive. Finally, in quadrant IV, "Class," only cosine is positive.

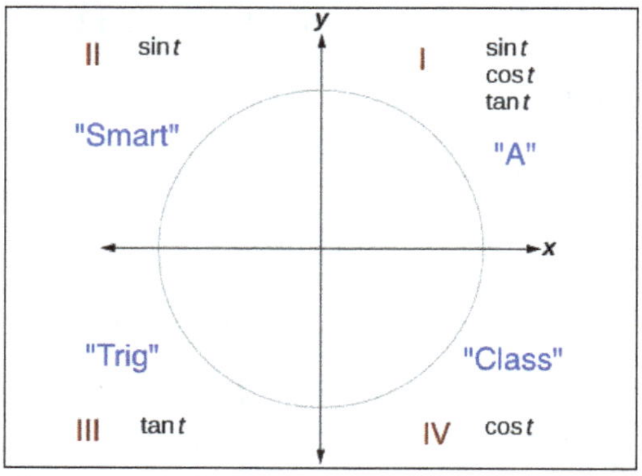

Sign rules for trigonometric functions: The trigonometric functions are each listed in the quadrants in which they are positive.

Identifying Values using Reference Angles

Take a close look at the unit circle, and note that sin t and cos t take certain values as they fluctuate between −1 and 1. You will notice that they take on the value of zero, as well as the positive and negative values of three particular numbers: $\frac{\sqrt{3}}{2}, \frac{\sqrt{2}}{2}$ and $\frac{1}{2}$.

Identifying reference angles will help us identify a pattern in these values.

Reference angles in quadrant I are used to identify which value any angle in quadrants II, III, or IV will take. This means that we only need to memorize the sine and cosine of three angles in quadrant I: 30°, 45°, and 60°.

For any given angle in the first quadrant, there is an angle in the second quadrant with the same sine value. Because the sine value is the y-coordinate on the unit circle, the other angle with the same sine will share the same y-value, but have the opposite x-value. Therefore, its cosine value will be the opposite of the first angle's cosine value.

Likewise, there will be an angle in the fourth quadrant with the same cosine as the original angle. The angle with the same cosine will share the same x-value but will have the opposite y-value. Therefore, its sine value will be the opposite of the original angle's sine value.

As shown in the diagrams below, angle α has the same sine value as angle t; the cosine values are opposites. Angle β has the same cosine value as angle t; the sine values are opposites.

$$\sin t = \sin\alpha \quad \text{and} \quad \cos t = -\cos\alpha$$
$$\sin t = -\sin\beta \quad \text{and} \quad \cos t = \cos\beta$$

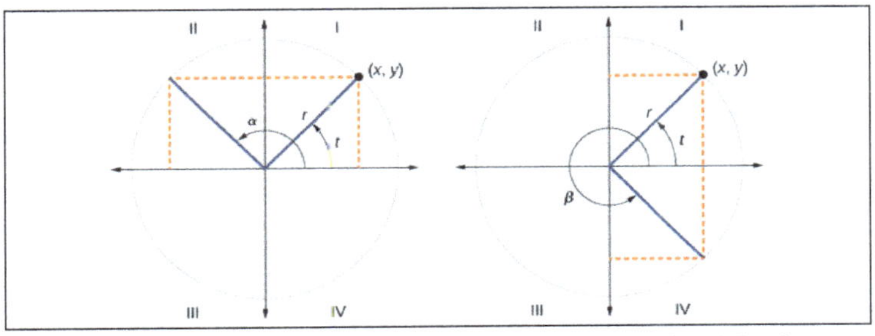

Reference angles: In the left figure, t is the reference angle for α. In the right figure, t is the reference angle for β.

Recall that an angle's reference angle is the acute angle, t, formed by the terminal side of the angle t and the horizontal axis. A reference angle is always an angle between 0 and 90°, or 0 and $\frac{\pi}{2}$ radians. For any angle in quadrants II, III, or IV, there is a reference angle in quadrant I.

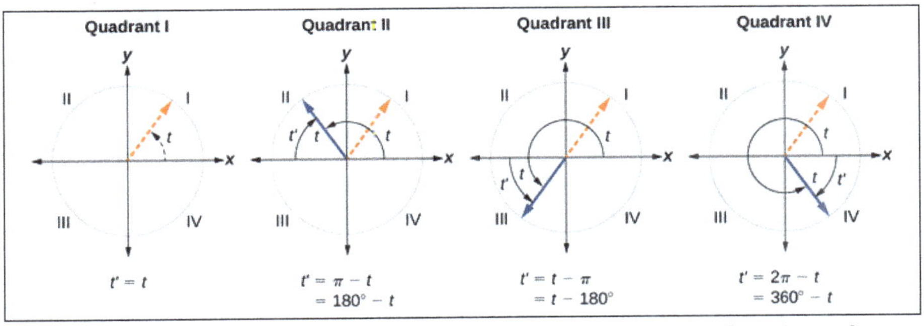

Reference angles in each quadrant: For any angle in quadrants II, III, or IV, there is a reference angle in quadrant I.

Thus, in order to recall any sine or cosine of a special angle, you need to be able to identify its angle with the x-axis in order to compare it to a reference angle. You will then identify and apply the appropriate sign for that trigonometric function in that quadrant.

These are the steps for finding a reference angle for any angle between 0 and 2π:

- An angle in the first quadrant is its own reference angle.
- For an angle in the second or third quadrant, the reference angle is $|\pi - t|$ or $|180° - t|$.
- For an angle in the fourth quadrant, the reference angle is 2π−t or 360° − t. If an angle is less than 0 or greater than 2π, add or subtract 2π as many times as needed to find an equivalent angle between 0 and 2π.

Since tangent functions are derived from sine and cosine, the tangent can be calculated for any of the special angles by first finding the values for sine or cosine.

Example: Find $\tan(225^\circ)$, applying the rules above.

First, note that 225° falls in the third quadrant:

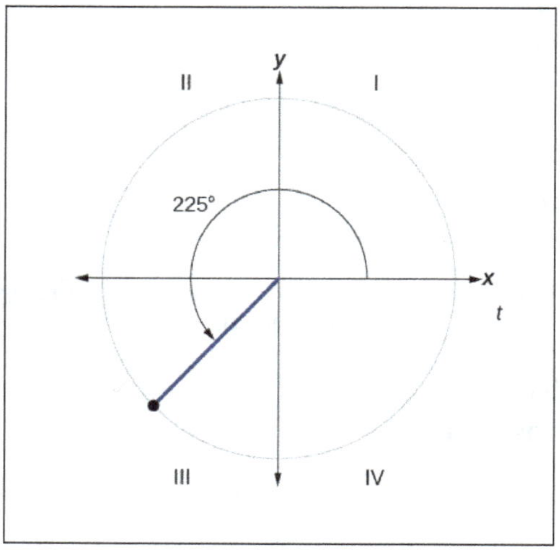

Angle 225° on a unit circle: The angle 225° falls in quadrant III.

Subtract 225° from 180° to identify the reference angle:

$$|180^\circ - 225^\circ| = |-45^\circ|$$
$$= 45^\circ$$

In other words, 225° falls 45° from the x-axis. The reference angle is 45°.

Recall that,

$$\sin(45^\circ) = \frac{\sqrt{2}}{2}$$

However, the rules described above tell us that the sine of an angle in the third quadrant is negative. So we have,

$$\sin(225^\circ) = -\frac{\sqrt{2}}{2}$$

Following the same process for cosine, we can identify that,

$$\cos(225^\circ) = \frac{\sqrt{2}}{2}$$

We can find $\tan(225^\circ)$ by dividing $\sin(225^\circ)$ by $\cos(225^\circ)$:

$$\begin{aligned}\tan(225^\circ) &= \frac{\sin(225^\circ)}{\cos(225^\circ)} \\ &= \frac{-\frac{\sqrt{2}}{2}}{-\frac{\sqrt{2}}{2}} \\ &= -\frac{\sqrt{2}}{2} \cdot -\frac{2}{\sqrt{2}} \\ &= 1\end{aligned}$$

References

- Sine: mathworld.wolfram.com, Retrieved 02 June, 2019
- Sine-Function-is-Odd: proofwiki.org, Retrieved 28 April, 2019
- Cosine: mathworld.wolfram.com, Retrieved 14 May, 2019
- Trigonometric-functions-and-the-unit-circle, boundless-algebra: courses.lumenlearning.com, Retrieved 19 August, 2019
- Derivative-of-Cosecant-Function: proofwiki.org, Retrieved 27 March, 2019
- Tangent: mathworld.wolfram.com, Retrieved 26 January, 2019

3

Inverse Trigonometric Functions

Inverse trigonometric functions are the inverse functions of sine, cosine, tangent, cotangent, secant, and cosecant functions. They can be used for obtaining an angle from any of the angle's trigonometric ratios. All these inverse trigonometric functions have been carefully analyzed in this chapter.

Inverse Sine

The inverse sine is the multivalued function $\sin^{-1} z$, also denoted $\arcsin z$, that is the inverse function of the sine. The variants $\text{Arcsin}\, z$ and $\text{Sin}^{-1} z$ are sometimes used to refer to explicit principal values of the inverse sine, although this distinction is not always made. Worse yet, the notation $\arcsin z$ is sometimes used for the principal value, with $\text{Arcsin}\, z$ being used for the multivalued function. Note that in the notation $\sin^{-1} z$ (commonly used in North America and in pocket calculators worldwide), $\sin z$ is the sine and the superscript −1 denotes the inverse function, *not* the multiplicative inverse.

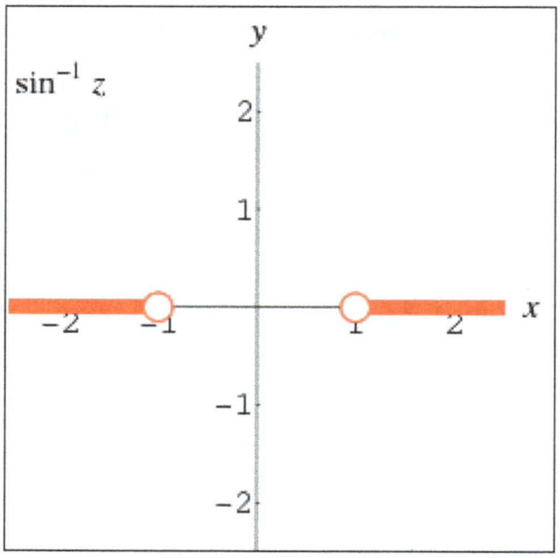

The inverse sine is a multivalued function and hence requires a branch cut in the complex plane. This follows from the definition of $\sin^{-1} z$ as:

$$sin^{-1} z = -i \ln\left(i z + \sqrt{1 - z^2}\right).$$

Special values include:

$$\sin^{-1}(-1) = -\frac{1}{2}\pi$$

$$\sin^{-1} 0 = 0$$

$$\sin^{-1} 1 = \frac{1}{2}\pi.$$

The derivative of $\sin^{-1} z$ is:

$$\frac{d}{dz}\sin^{-1} z = \frac{1}{\sqrt{1-z^2}}$$

and its indefinite integral is:

$$\frac{d}{dz}\sin^{-1} z = \frac{1}{\sqrt{1-z^2}}$$

and its indefinite integral is:

$$\int \sin^{-1} z \, d\,z = \sqrt{1-z^2} + z\sin^{-1} z + C.$$

The inverse sine satisfies:

$$= \sin^{-1} z = \csc^{-1}\left(\frac{1}{z}\right)$$

For $z \neq 0$,

$$\begin{aligned} \sin^{-1} z &= -\sin^{-1}(-z) \\ &= \cos^{-1}(-z) - \frac{1}{2}\pi \\ &= \frac{1}{2}\pi - \cos^{-1} z \end{aligned}$$

for all complex z,

$$\sin^{-1} x = \begin{cases} -\frac{1}{2}\pi + \sin^{-1}\left(\sqrt{1-x^2}\right) & \text{for } x < 0 \\ \frac{1}{2}\pi - \sin^{-1}\left(\sqrt{1-x^2}\right) & \text{for } x > 0 \end{cases}$$

$$= \begin{cases} -\frac{1}{2}\pi - \cot^{-1}\left(\frac{x}{\sqrt{1-x^2}}\right) & \text{for } x<0 \\ \frac{1}{2}\pi - \cot^{-1}\left(\frac{x}{\sqrt{1-x^2}}\right) & \text{for } x>0 \end{cases}$$

$$= \begin{cases} -\cos^{-1}\left(\sqrt{1-x^2}\right) & \text{for } 1-<x<0 \\ \cos^{-1}\left(\sqrt{1-x^2}\right) & \text{for } 0<x<1 \end{cases}$$

$$= \begin{cases} -\sec^{-1}\left(\frac{1}{\sqrt{1-x^2}}\right) & \text{for } -1<x<0 \\ \sec^{-1}\left(\frac{1}{\sqrt{1-x^2}}\right) & \text{for } 0<x<1, \end{cases}$$

And,

$$\sin^{-1} x = \tan^{-1}\left(\frac{x}{\sqrt{1-x^2}}\right)$$

$$= \cot^{-1}\left(\frac{\sqrt{1-x^2}}{x}\right)$$

for $-1 < x < 1$, where equality at points where the denominators are 0 is understood to mean in the limit as $x \mapsto \pm 1$ or $x \to 0,$, respectively.

The Maclaurin series for the inverse sine with $-1 \leq \quad \leq 1$ is given by:

$$\sin^{-1} x = \sum_{n=0}^{\infty} \frac{\left(\frac{1}{2}\right)_n}{(2n+1)n!} x^{2n+1}$$

$$= x + \frac{1}{6}x^3 + \frac{3}{40}x^5 + \frac{5}{112}x^7 + \frac{35}{1152}x^9 + \ldots$$

Where $\binom{2n}{n}$ is a binomial coefficient. Similarly,

$$\left[\sin^{-1}\left(\frac{1}{2}x\right)\right]^4 = \frac{3}{2}\sum_{k=1}^{\infty}\left[\sum_{m=1}^{k-1}\frac{1}{m^2}\right]\frac{x^{2k}}{k^2\binom{2k}{k}}$$

$$\left[\sin^{-1}\left(\frac{1}{2}x\right)\right]^6=\frac{45}{4}\sum_{k=1}^{\infty}\left[\sum_{m=1}^{k-1}\frac{1}{m^2}\sum_{n=1}^{m-1}\frac{1}{n^2}\right]\frac{x^{2k}}{k^2\binom{2k}{k}}$$

$$\left[\sin^{-1}\left(\frac{1}{2}x\right)\right]^8=\frac{315}{2}\sum_{k=1}^{\infty}\left[\sum_{m=1}^{k-1}\frac{1}{m^2}\sum_{n=1}^{m-1}\frac{1}{n^2}\sum_{p=1}^{n-1}\frac{1}{p^2}\right]\frac{x^{2k}}{k^2\binom{2k}{k}}$$

Ramanujan gave the cases $\left(\sin^{-1}x\right)^n$ for $n=1$, 2, 3, and 4, and the general cases are given in terms of multiple sums by Bailey *et al.* and Borwein and Chamberland.

The inverse sine has continued fraction:

$$\sin^{-1}z=\cfrac{z\sqrt{1-z^2}}{1-\cfrac{1\cdot 2z^2}{3-\cfrac{1\cdot 2z^2}{5-\cfrac{3\cdot 4z^2}{7-\cfrac{3\cdot 4z^2}{9-\cfrac{5\cdot 6z^2}{11-\ldots}}}}}}$$

Arcsin Rules

Rule name	Rule
Sine of arcsine	$\sin(\arcsin x)=x$
Arcsine of sine	$\arcsin(\sin x)=x+2k\pi$, when $k\in\mathbb{Z}$ (k is integer)
Arcsin of negative argument	$\arcsin(-x)=-\arcsin x$
Complementary angles	$\arcsin x=\pi/2-\arccos x=90°-\arccos x$
Arcsin sum	$\arcsin\alpha+\arcsin(\beta)=\arcsin\left(\alpha\sqrt{(1-\beta^2)}+\beta\sqrt{(1-\alpha^2)}\right)$
Arcsin difference	$\arcsin\alpha-\arcsin(\beta)=\arcsin\left(\alpha\sqrt{(1-\beta^2)}-\beta\sqrt{(1-\alpha^2)}\right)$
Cosine of arcsine	$\cos(\arcsin x)=\sin(\arccos x)=\sqrt{1-x^2}$
Tangent of arcsine	$\tan(\arcsin x)=\dfrac{x}{\sqrt{1-x^2}}$

Derivative of arcsine	$\frac{d}{dx}(\arcsin x) = \arcsin' x = \frac{1}{\sqrt{1-x^2}}$
Indefinite integral of arcsine	$\int \arcsin x\, dx = x \arcsin x + \sqrt{1-x^2} + C$

Arcsin Table

x	$\arcsin(x)$ (rad)	$\arcsin(x)$ $(°)$
−1	$-\pi/2$	-90°
$-\sqrt{3/2}$	$-\pi/3$	-60°
$-\sqrt{2/2}$	$-\pi/4$	-45°
−1/2	$-\pi/6$	-30°
0	0	0°
1/2	$\pi/6$	30°
√2/2	$\pi/4$	45°
√3/2	$\pi/3$	60°
1	$\pi/2$	90°

Inverse Cosine

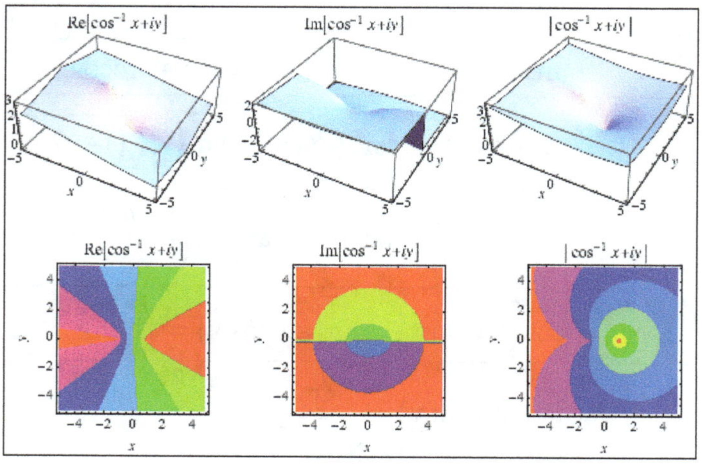

The inverse cosine is the multivalued function $\cos^{-1} z$, also denoted $\arccos z$, that is the inverse function of the cosine. The variants $\operatorname{Arccos} z$ and $\operatorname{Cos}^{-1} z$ are sometimes used to refer to explicit principal values of the inverse cosine, although this distinction is not always made. Worse yet, the notation $\arccos z$ is sometimes used for the principal value, with $\operatorname{Arccos} z$ being used for the multivalued function Note that the notation $\cos^{-1} z$ (commonly used in North America and in pocket calculators worldwide), $\cos z$ is the cosine and the superscript −1 denotes the inverse function, not the multiplicative inverse.

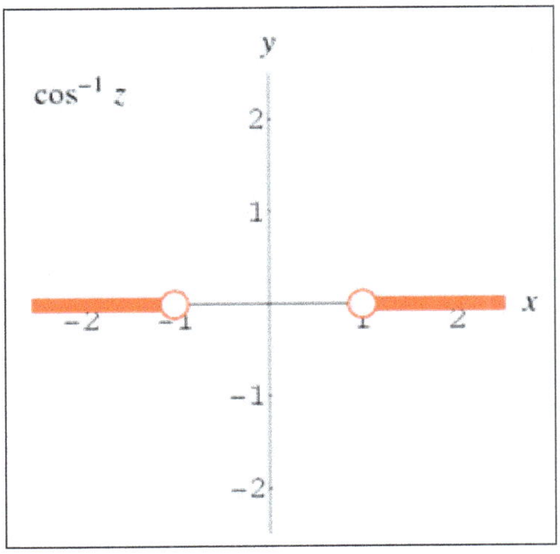

The inverse cosine is a multivalued function and hence requires a branch cut in the complex plane. This follows from the definition of $\cos^{-1} z$ as:

$$\cos^{-1} z = \frac{1}{2}\pi + i \operatorname{In}\left(i z + \sqrt{1+z^2}\right).$$

Special values include:

$$\cos^{-1}(-1) = \pi$$

$$\cos^{-1} 0 = \frac{1}{2}\pi$$

$$\cos^{-1} 1 = 0.$$

The derivative of $\cos^{-1} z$ is given by:

$$\frac{d}{dz}\cos^{-1} z = \frac{1}{\sqrt{1-z^2}}$$

and its indefinite integral is:

$$\int \cos^{-1} z \, d\, z = z\cos^{-1} z - \sqrt{1-z^2} + C.$$

The inverse cosine satisfies:

$$\cos^{-1} z = \pi - \cos^{-1}(-z)$$

for all complex z, and:

$$\cos^{-1} x = \begin{cases} \frac{1}{2}\pi + \cos^{-1}\left(\sqrt{1-x^2}\right) & \text{for } x \leq 0 \\ \frac{1}{2}\pi - \cos^{-1}\left(\sqrt{1-x^2}\right) & \text{for } x \geq 0 \end{cases}$$

The inverse cosine is given in terms of other inverse trigonometric functions by:

$$\begin{aligned} \cos^{-1} z &= \frac{1}{2}\pi + \sin^{-1}(-z) \\ &= \frac{1}{2}\pi - \sin^{-1} z \end{aligned}$$

for all complex z,

$$\cos^{-1} z = \sec^{-1}\left(\frac{1}{z}\right)$$

for $z \neq 0$,

$$\cos^{-1} x = \frac{1}{2}\pi - \tan^{-1}\left(\frac{x}{\sqrt{1-x^2}}\right)$$

for $-1 \leq x \leq 1$, and:

$$\begin{aligned} \cos^{-1} x &= \cot^{-1}\left(\frac{x}{\sqrt{1-x^2}}\right) \\ &= \csc^{-1}\left(\frac{1}{\sqrt{1-x^2}}\right) \\ &= \sin^{-1}\left(\sqrt{1-x^2}\right) \\ &= \tan^{-1}\left(\frac{\sqrt{1-x^2}}{x}\right) \end{aligned}$$

for $x \geq 0$, where in the last equation, equality at zero is understood to mean in the limit as $x \to 0^+$.

The Maclaurin series for the inverse cosine with $-1 \le x \le 1$ is:

$$\cos^{-1} x = \frac{1}{2}\pi - \sum_{n=1}^{\infty} \frac{\binom{\frac{1}{2}}{n-1}}{(n-1)!(2n-1)} x^{2n-1}$$

$$= \frac{1}{2}\pi - x - \frac{1}{6}x^3 - \frac{3}{40}x^5 - \frac{5}{112}x^7 - \frac{35}{1152}x^9 - \ldots$$

Arccos Rules

Rule name	Rule
Cosine of arccosine	$\cos(\arccos x) = x$
Arccosine of cosine	$\arccos(\cos x) = x + 2k\pi$, when $k \in \mathbb{Z}$ (k is integer)
Arccos of negative argument	$\arccos(-x) = \pi - \arccos x = 180° - \arccos x$
Complementary angles	$\arccos x = \pi / 2 - \arcsin x = 90° - \arcsin x$
Arccos sum	$\arccos(\alpha) + \arccos(\beta) = \arccos\left(\alpha\beta - \sqrt{(1-\alpha^2)(1-\beta^2)}\right)$
Arccos difference	$\arccos(\alpha) - \arccos(\beta) = \arccos\left(\alpha\beta + \sqrt{(1-\alpha^2)(1-\beta^2)}\right)$
Arccos of sin of x	$\arccos(\sin x) = -x - (2k + 0.5)\pi$
Sine of arccosine	$\cos(\arcsin x) = \sin(\arccos x) = \sqrt{1-x^2}$
Tangent of arccosine	$\tan(\arccos x) = \frac{\sqrt{1-x^2}}{x}$
Derivative of arccosine	$\frac{d}{dx}(\arccos x) = \arccos' x = \frac{-1}{\sqrt{1-x^2}}$
Indefinite integral of arccosine	$\int \arccos x \, dx = x \arccos x - \sqrt{1-x^2} + C$

Arccos Table

x	$\arccos(x)$ (rad)	$\arccos(x)$ (°)

−1	π	180°
$-\sqrt{3/2}$	$5\pi/6$	150°
$-\sqrt{2/2}$	$3\pi/4$	135°
− 1/2	$2\pi/3$	120°
0	$\pi/2$	90°
1/2	$\pi/3$	60°
$\sqrt{2/2}$	$\pi/4$	45°
$\sqrt{3/2}$	/6	30°
1	0	0°

Inverse Tangent

The inverse tangent is the multivalued function $\tan^{-1} z$, also denoted $\arctan z$ or $\text{arctg}\, z$, that is the inverse function of the tangent. The variants $\text{Arctan}\, z$ and $\text{Tan}^{-1} z$ are sometimes used to refer to explicit principal values of the inverse cotangent, although this distinction is not always made.

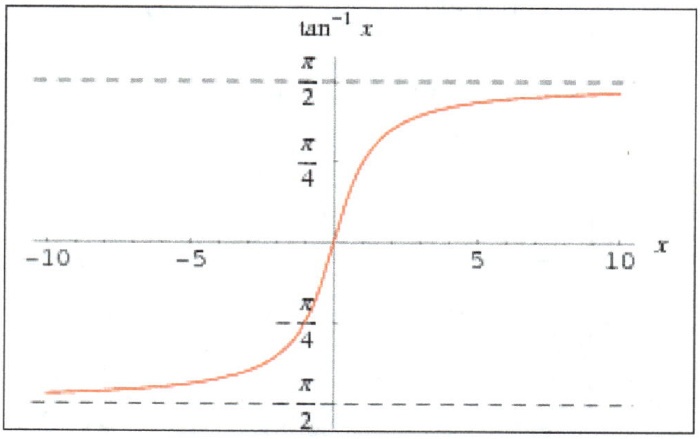

The inverse tangent function $\tan^{-1} x$ is plotted above along the real axis.

Worse yet, the notation $\arctan z$ is sometimes used for the principal value, with $\text{Arctan}\, z$ being used for the multivalued function. Note that in the notation $\tan^{-1} z$ (commonly used in North America and in pocket calculators worldwide), $\tan z$ denotes the tangent and −1 the inverse function, not the multiplicative inverse.

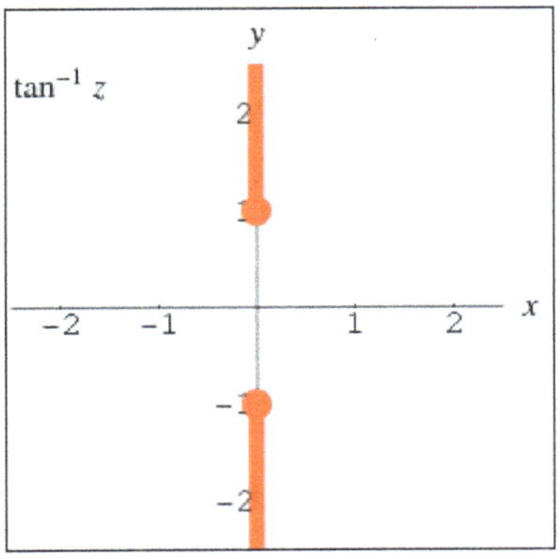

The inverse tangent is a multivalued function and hence requires a branch cut in the complex plane. This follows from the definition of $\tan^{-1} z$ as:

$$\tan^{-1} z = \frac{1}{2} i \left[\text{In}(1 - i z) - \text{In}(1 + i z) \right].$$

This branch cut definition determines the range of $\tan^{-1} x$ for real x as $(-\pi/2, \pi/2)$. Care must be taken, however, as other branch cut definitions can give different ranges (most commonly, $(0, \pi)$).

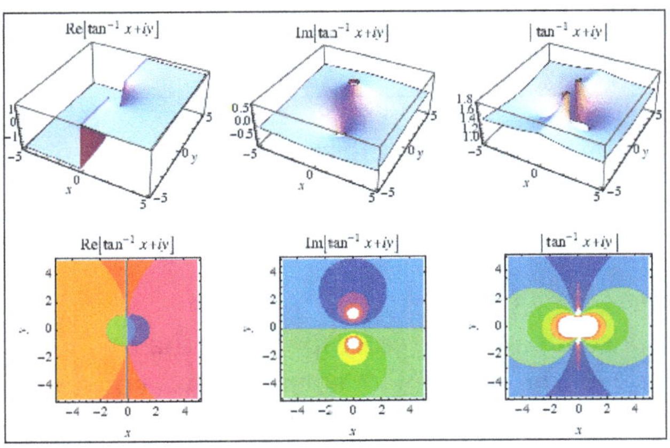

The inverse tangent function $\tan^{-1} z$ is plotted above in the complex plane.

$\tan^{-1} z$ has the special values:

$$\tan^{-1}(-\infty) = -\frac{1}{2}\pi$$

$$\tan^{-1}(-i) = -i\infty$$

$$\tan^{-1} 0 = 0$$

$$\tan^{-1} i = i\infty$$

$$\tan^{-1} \infty = \frac{1}{2}\pi.$$

The derivative of $\tan^{-1} z$ is:

$$\frac{d}{dz}\tan^{-1} z = \frac{1}{1+z^2}$$

and the indefinite integral is:

$$\int \tan^{-1} z \, d\, z = z \tan^{-1} z - \frac{1}{2}\text{In}\left(1+z^2\right)+C.$$

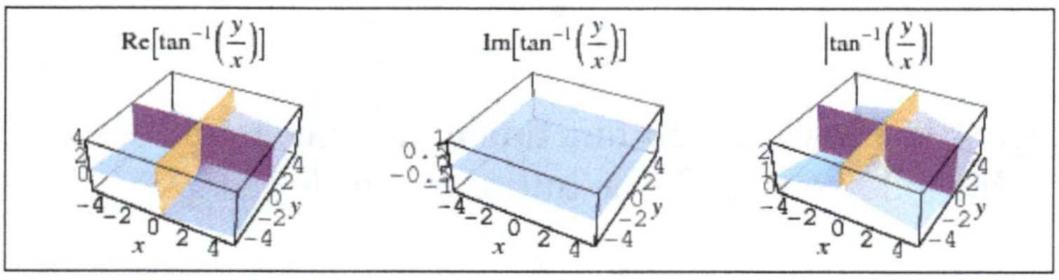

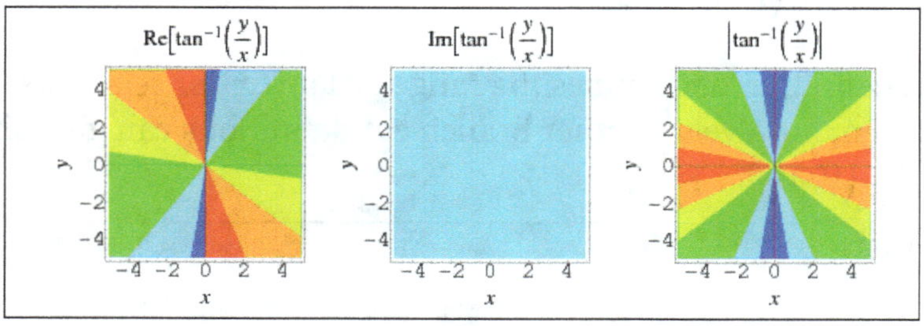

The complex argument of a complex number $z = x + i\,y$ is often written as:

$$\theta = \tan^{-1}\left(\frac{y}{x}\right),$$

where θ, sometimes also denoted ϕ, corresponds to the counterclockwise angle from the positive real axis, i.e., the value of θ such that $x = \cos\theta$ and $y = \sin\theta$. Plots of $\tan^{-1}(y/x)$ are illustrated above for real values of x and y.

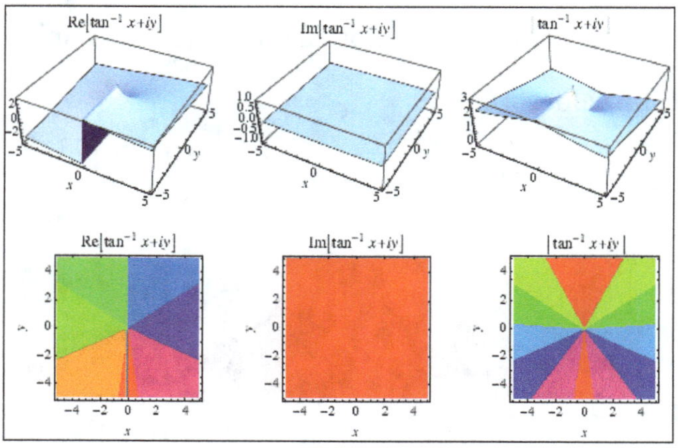

In the degenerate case when $x = 0$,

$$\phi = \begin{cases} -\frac{1}{2}\pi & \text{if } y < 0 \\ \text{undefined} & \text{if } y = 0 \\ \frac{1}{2}\pi & \text{if } y > 0 \end{cases}$$

The usual $\tan^{-1} z$ has the Maclaurin series of:

$$\tan^{-1} z = \sum_{n=0}^{\infty} \frac{(-1)^n z^{2n+1}}{2n+1}$$
$$= z - \frac{1}{3}z^3 + \frac{1}{5}z^5 - \frac{1}{7}z^7 + \ldots$$

A more rapidly converging form due to Euler is given by:

$$\tan^{-1} x = \sum_{n=0}^{\infty} \frac{2^{2n}(n!)^2}{(2n+1)!} \frac{x^{2n+1}}{(1+x^2)^{n+1}}$$

for real x. This is related to the formula of Euler given by:

$$\tan^{-1} x \frac{y}{x}\left(1 + \frac{2}{3}y + \frac{2\cdot 4}{3\cdot 5}y^2 + \frac{2\cdot 4\cdot 6}{3\cdot 6\cdot 7}y^3 + \ldots.\right),$$

Where:

$$y \equiv \frac{x^2}{1+x^2}.$$

The inverse tangent formulas are connected with many interesting approximations to pi:

$$\tan^{-1}(1+x) = \frac{\pi}{4} + i\sum_{n=1}^{\infty} \frac{(-1-i)^n - (i-1)^n}{2^{n+1} n} x^n$$
$$= \frac{1}{4}\pi + \frac{1}{2}x - \frac{1}{4}x^2 + \frac{1}{12}x^3 - \frac{1}{40}x^5 + \frac{1}{48}x^6 - \frac{1}{112}x^7 + \ldots$$

The inverse tangent satisfies:

$$\tan^{-1} z = -\cot^{-1}\left(\frac{1}{z}\right)$$

for $z \neq 0$,

$$\tan^{-1} z = -\tan^{-1}(-z)$$

for all complex z,

$$\tan^{-1} x = \frac{1}{2}\pi - \cos^{-1}\left(\frac{x}{\sqrt{x^2+1}}\right)$$

$$= \sin^{-1}\left(\frac{x}{\sqrt{x^2+1}}\right)$$

$$= \csc^{-1}\left(\frac{\sqrt{x^2+1}}{x}\right)$$

for all real x, where equality for the last equation is understood to be in the limit as $x \to 0$, and:

$$\tan \quad \begin{cases} -\frac{1}{\ } \quad -\tan \left(\frac{1}{\ }\right) & \text{for } > 0 \\ \frac{1}{\ } \quad -\tan \left(\frac{1}{\ }\right) & \text{for } > 0 \end{cases}$$

$$\begin{cases} -- \quad +\cot \ (-x) & \text{for } x < 0 \\ - \quad +\cot \ (-x) & \text{for } x > 0 \end{cases}$$

$$\begin{cases} -- \quad -\cot \ x & \text{for } x > 0 \\ - \quad -\cot \ x & \text{for } x > 0 \end{cases}$$

$$\begin{cases} -\cos \left(\frac{\ }{\sqrt{\ }}\right) & \text{for } < 0 \\ \cos \left(\frac{\ }{\sqrt{\ }}\right) & \text{for } \quad 0 \end{cases}$$

$$\begin{cases} -\sec^{\ 1}\left(\sqrt{x^2+1}\right) & \text{for } x < 0 \\ \sec^{\ 1}\left(\sqrt{x^2+1}\right) & \text{for } x > 0 \end{cases}$$

In terms of the hypergeometric function,

$$\tan^{-1} z = z_2 F_1\left(1, \frac{1}{2}, \frac{3}{2}; -z^2\right)$$

for complex z, and:

$$\tan^{-1} x = \frac{x}{1+x^2} 2F_1\left(1,1; \frac{3}{2}; \frac{x^2}{1+x^2}\right)$$

for real x.

Castellanos also gives some curious formulas in terms of the Fibonacci numbers,

$$\tan^{-1} x = \sum_{n=0}^{\infty} \frac{(-1)^n F_{2n+1} t^{2n+1}}{5^n (2n+1)}$$

$$= 5\sum_{n=0}^{\infty} \frac{(-1)^n F_{2n+1}^2}{(2n+1)\left(u+\sqrt{u_2+1}\right)^{2n+1}}$$

$$= \sum_{n=0}^{\infty} \frac{(-1)^n 5^{n+2} F_{2n+1}^3}{(2n+1)\left(v+\sqrt{v^2+5}\right)^{2n+1}},$$

Where:

$$t \equiv \frac{2x}{1+\sqrt{1+\frac{4x^2}{5}}}$$

$$u \equiv \frac{5}{4x}\left(1+\sqrt{1+\frac{24}{25}x^2}\right),$$

and v is the largest positive root of:

$$8xv^4 - 100v^3 - 450xv^2 + 875v + 625x = 0.$$

The inverse tangent satisfies the addition formula:

$$\tan^{-1} x + \tan^{-1} y = \tan^{-1}\left(\frac{x \quad y}{x\, y}\right)$$

for $-1 < x, y < 1$, as well as the more complicated formula:

$$\tan^{-1}\left(\frac{1}{a}\right) = 2\tan^{-1}\left(\frac{1}{2a}\right) - \tan^{-1}\left(\frac{1}{4a^3 + 3a}\right)$$

valid for all complex a. An additional identity known to Euler is given by:

$$\tan^{-1}\left(\frac{1}{a-b}\right) = \tan^{-1}\left(\frac{1}{a}\right) + \tan^{-1}\left(\frac{b}{a^2 - ab + 1}\right)$$

for $(a > b \wedge a > 0)$ or $(a < b \wedge a < 0)$. Another interesting inverse tangent identity attributed to Charles Dodgson (Lewis Carroll) by Lehmer is:

$$\tan^{-1}(p+r) + \tan^{-1}(p+q) - \tan^{-1} p = \frac{1}{2}\pi,$$

Where:

$$1 + p^2 = qr$$

and $p, q, r > 0$

The inverse tangent has continued fraction representations:

$$\tan^{-1} x = \cfrac{x}{1 + \cfrac{x^2}{3 + \cfrac{4x^2}{5 + \cfrac{9x^2}{7 + \cfrac{16x^2}{9 + \dots}}}}}$$

and:

$$\tan^{-1} x = \cfrac{x}{1 + \cfrac{x^2}{3 - x^2 + \cfrac{9x^2}{5 - 3x^2 \cfrac{25x^2}{7 - 5x^2 + \dots}}}}$$

due to Euler and sometimes known as Euler's continued fraction.

To find $\tan^{-1} x$ numerically, the following arithmetic-geometric mean-like algorithm can be used. Let:

$$a_0 = \left(1+x^2\right)^{-1/2}$$
$$b_0 = 1.$$

Then compute:

$$a_{i+1} = \frac{1}{2}\left(a_j + b_i\right)$$
$$b_{j+1} = \sqrt{a_{i+1} b_i},$$

and the inverse tangent is given by:

$$\tan^{-1} x = \lim_{n\to\infty} \frac{x}{\sqrt{1+x^2}\ \ a_n}$$

An inverse tangent $\tan^{-1} n$ with integral n is called reducible if it is expressible as a finite sum of the form:

$$\tan^{-1} n = \sum_{k=1} f_k \tan^{-1} n_k,$$

where f_k are positive or negative integers and are integers $< n$. $\tan^{-1} m$ is reducible iff all the prime factors of $1+m^2$ occur among the prime factors of $1+n^2$ for $n = 1$, ..., $m-1$. A second necessary and sufficient condition is that the largest prime factor of $1+m^2$ is less than $2m$. Equivalent to the second condition is the statement that every Gregory number $t_x = \cot^{-1} x$ can be uniquely expressed as a sum in terms of t_ms for which m is a Størmer number. To find this decomposition, write:

$$\arg(1+in) = \arg \prod_{k=1} (1+n_k i)^{f_k},$$

so the ratio:

$$r = \frac{\Pi_{k=1}(1+n_k i)^{f_k}}{1+in}$$

is a rational number. Equation $r = \frac{\Pi_{k=1}(1+n_k i)^{f_k}}{1+in}$ can also be written:

$$r^2\left(1+n^2\right) = \prod_{k=1}\left(1+n_k^2\right)^{f_k}.$$

Writing in the form:

$$\tan^{-1} n = \sum_{k=1} f_k \tan^{-1} n_k + f \tan^{-1} 1$$

allows a direct conversion to a corresponding inverse cotangent formula:

$$\cot^{-1} n = \sum_{k=1} f_k \cot^{-1} n_k + c \cot^{-1} 1,$$

Where:

$$c = 2 - f - 2\sum_{k=1} f_r.$$

Todd gives a table of decompositions of $\tan^{-1} n$ for $n \leq 342$. Conway and Guy give a similar table in terms of Størmer numbers.

Arndt and Gosper give the remarkable inverse tangent identity:

$$\sin\left(\sum_{k=1}^{2n+1} \tan^{-1} a_k\right) = \frac{(-1)^n}{2n+1} \frac{\Sigma_{k=1}^{2n+1} \Pi_{j=1}^{2n+1}\left[a_j - \tan\left(\frac{\pi(j-k)}{2n+1}\right)\right]}{\sqrt{\Pi_{j=1}^{2n+1}(a_j^2+1)}}.$$

There is an amazing set of BBP-type formulas for $\tan^{-1}(4/5)$:

$$\tan^{-1}\left(\frac{4}{5}\right) = \begin{array}{l} \frac{1}{131072}\sum_{k=0}^{\infty}\frac{1}{1048576^k}\left[\frac{262144}{40k+2} - \frac{163840}{40k+5} - \frac{65536}{40k+6}\right. + \\ \frac{16384}{40k+10} - \frac{4096}{40k+14} - \frac{5120}{40k+15} + \frac{1024}{40k+18} - \frac{256}{40k+22} + \\ \left.\frac{160}{40k+25} + \frac{64}{40k+26} - \frac{16}{40k+30} + \frac{4}{40k+34} + \frac{5}{40k+35} - \frac{1}{40k+34}\right] \end{array}$$

$$\begin{array}{l} \frac{1}{131072}\sum_{k=0}^{\infty}\frac{1}{1048576^k}\left[\frac{393216}{40k+4} + \frac{163840}{40k+5} - \frac{131072}{40k+6} - \frac{163840}{40k+8} + \frac{24576}{40k+12}\right. - \\ = \frac{8192}{40k+14} - \frac{15360}{40k+15} - \frac{10240}{40k+16} - \frac{1024}{40k+20} - \frac{512}{40k-6} - \frac{640}{40k+24} - \\ \left.\frac{160}{40k+25} + \frac{96}{40k+28} - \frac{32}{40k+30} - \frac{40}{40k+32} + \frac{15}{40k+35} + \frac{6}{40k+36} - \frac{2}{40k+38}\right] \end{array}$$

$$\frac{1}{131072}\sum_{k=0}^{\infty}\frac{1}{1048576^k}\left[\frac{262144}{40k+1}-\frac{262144}{40k+3}-\frac{65536}{40k+5}-\frac{327680}{40k+6}+\frac{65536}{40k+7}-\right.$$
$$\frac{163840}{40k+8}+\frac{16384}{40k+9}-\frac{40960}{40k+10}-\frac{16384}{40k+11}-\frac{4096}{40k+13}-\frac{20480}{40k+14}-\frac{16384}{40k+15}-$$
$$=\frac{10240}{40k+16}+\frac{1024}{40k+17}-\frac{1024}{40k+19}-\frac{2560}{40k+20}-\frac{256}{40k+21}-\frac{1280}{40k+22}+$$
$$\frac{256}{40k+23}-\frac{640}{40k+24}+\frac{64}{40k+25}-\frac{64}{40k+27}-\frac{16}{40k+29}-\frac{40}{40k+30}+$$
$$\left.\frac{16}{40k+31}-\frac{40}{40k+32}+\frac{4}{40k+33}+\frac{16}{40k+35}-\frac{1}{40k+37}-\frac{5}{40k+38}+\frac{1}{40k+39}\right]$$

$$\frac{1}{262144}\sum_{k=0}^{\infty}\frac{1}{1048576^k}\left[\frac{262144}{40k+3}+\frac{262144}{40k+4}+\frac{131072}{40k+6}-\frac{65536}{40k+7}+\right.$$
$$=\frac{81920}{40k+10}+\frac{16384}{40k+11}+\frac{16384}{40k+12}+\frac{8192}{40k+14}-\frac{4096}{40k+15}+$$
$$\frac{1024}{40k+19}+\frac{1024}{40k+20}+\frac{512}{40k+22}-\frac{256}{40k+23}+\frac{64}{40k+27}+\frac{64}{40k+28}-$$
$$\left.\frac{48}{40k+30}-\frac{16}{40k+31}+\frac{4}{40k+35}+\frac{4}{40k+36}+\frac{2}{40k+38}-\frac{1}{40k+39}\right]$$

Arctan Rules

Rule name	Rule
Tangent of arctangent	$\tan(\arctan x)=x$
Arctan of negative argument	$\arctan(-x)=-\arctan x$
Arctan sum	$\arctan\alpha+\arctan\beta=\arctan\left[(\alpha+\beta)/(1-\alpha\beta)\right]$
Arctan difference	$\arctan\alpha-\arctan\beta=\arctan\left[(\alpha-\beta)/(1+\alpha\beta)\right]$
Sine of arctangent	$\sin(\arctan x)=\frac{x}{\sqrt{1+x^2}}$
Cosine of arctangent	$\cos(\arctan x)=\frac{1}{\sqrt{1+x^2}}$

Reciprocal argument	$\arctan\left(\frac{1}{x}\right)=\begin{cases}\frac{\pi}{2}-\arctan(x) & ,x>0\\ -\frac{\pi}{2}-\arctan(x) & ,x<0\end{cases}$
Arctan from arcsin	$\arctan x=\arcsin\frac{x}{\sqrt{x^2+1}}$
Derivative of arctan	$\frac{d}{dx}(\arctan x)=\arctan' x=\frac{1}{1+x^2}$
Indefinite integral of arctan	$\int \arctan x\,dx=x\arctan x-\frac{1}{2}\text{In }\left(1+x^2\right)+C$

Arctan Table

x	$\arctan(x)$ (rad)	$\arctan(x)$ $(°)$
$-\infty$	$-\pi/2$	-90°
−3	−1.2490	-71.565°
−2	−1.1071	-63.435°
$-\sqrt{3}$	$-\pi/3$	-60°
−1	$-\pi/4$	-45°
$-1/\sqrt{3}$	$-\pi/6$	-30°
−0.5	−0.4636	-26.565°
0	0	0°
0.5	0.4636	26.565°
$1/\sqrt{3}$	$\pi/6$	30°
1	$\pi/4$	45°
$\sqrt{3}$	$\pi/3$	60°

2	1.1071	63.435°
3	1.2490	71.565°
∞	$\pi/2$	90°

Inverse Cosecant

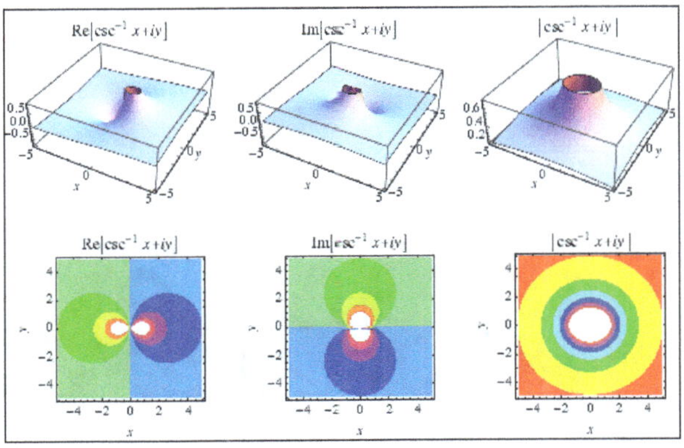

The inverse cosecant is the multivalued function $\csc^{-1} z$, also denoted $\text{arccsc}\, z$, that is the inverse function of the cosecant. The variants $\text{Arccsc}\, z$ and $\text{Csc}^{-1} z$ are sometimes used to refer to explicit principal values of the inverse cosecant, although this distinction is not always made. Worse yet, the notation $\text{arccsc}\, z$ is sometimes used for the principal value, with $\text{Arccsc}\, z$ being used for the multivalued function. Note that in the notation $\csc^{-1} z$ (commonly used in North America and in pocket calculators worldwide), $\csc z$ is the cosecant and the superscript –1 denotes an inverse function, not the multiplicative inverse.

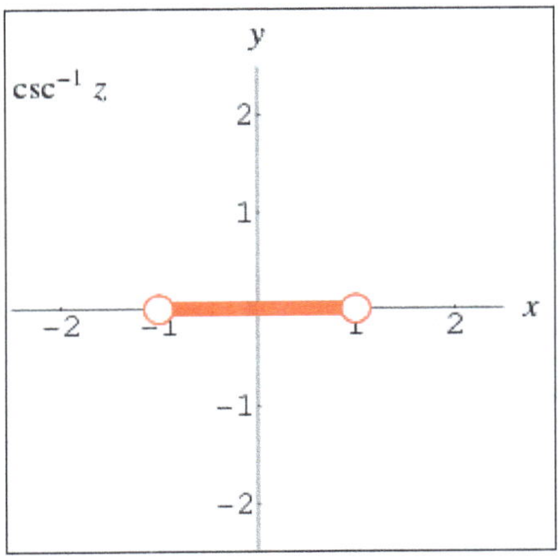

The inverse cosecant is a multivalued function and hence requires a branch cut in the complex plane. This follows from the definition of $\csc^{-1} z$ as:

$$\csc^{-1} z = -i \operatorname{In}\left(\sqrt{1-\frac{1}{z^2}}+\frac{i}{z}\right).$$

The derivative of $\csc^{-1} z$ is given by:

$$\frac{d}{dz}\csc^{-1} z = -\frac{1}{z^2\sqrt{1-\frac{1}{z^2}}},$$

which simplifies to:

$$\frac{d}{dx}\csc^{-1} x = -\frac{1}{x\sqrt{x^2-1}}$$

for x > 0. Its indefinite integral is:

$$\int \csc^{-1} z\, dz = z\csc^{-1} z + \operatorname{In}\left[z\left(1+\sqrt{\frac{z^2-1}{z^2}}\right)\right]+C$$

which simplifies to:

$$\int \csc^{-1} x\, dx = x\csc^{-1} x + \operatorname{In}\left(x+\sqrt{x^2-1}\right)$$

For x > 0.

The inverse cosecant has Taylor series about infinity of:

$$\csc^{-1} x = -\sum_{n=1}^{\infty}\frac{i^{n+1}P_{n-1}(0)}{n}x^{-n}$$

$$= \frac{\left(\frac{1}{2}\right)_{n-1}}{(n-1)!(2n-1)}x^{1-2}$$

$$= x^{-1}+\frac{1}{6}x^{-3}+\frac{3}{40}x^{-5}+\frac{5}{112}x^{-7}+\dots$$

Where $P_n(x)$ is a Legendre polynomial and $(x)_n$ is a Pochhammer symbol.

The inverse cosecant satisfies:

$$\csc^{-1} z = \sin^{-1}\left(\frac{1}{z}\right)$$

for $z \neq 0$,

$$\csc^{-1} z = \frac{1}{2}\pi - \sec^{-1} z$$
$$= -\frac{1}{2}\pi + \sec^{-1}(-z)$$

for all complex z, and

$$\csc^{-1} x = \begin{cases} \sec^{-1}\left(\frac{x}{\sqrt{x^2-1}}\right) - \pi & \text{for } x > -1 \\ \sec^{-1}\left(\frac{x}{\sqrt{x^2-1}}\right) & \text{for } x > 1 \end{cases}$$

$$= \begin{cases} -\cos^{-1}\left(\frac{\sqrt{x^2-1}}{x}\right) & \text{for } x < -1 \\ \cos^{-1}\left(\frac{\sqrt{x^2-1}}{x}\right) & \text{for } x > 1 \end{cases}$$

$$= \begin{cases} -\cot^{-1}\left(\sqrt{x^2-1}\right) & \text{for } x < -1 \\ \cot^{-1}\left(\sqrt{x^2-1}\right) & \text{for } x > 1. \end{cases}$$

Arccos Rules

Rule name	Rule
Cosine of arccosine	$\cos(\arccos x) = x$
Arccosine of cosine	$\arccos(\cos x) = x + 2k\pi$, when $k \in \mathbb{Z}$ (k is integer)
Arccos of negative argument	$\arccos(-x) = \pi - \arccos x = 180° - \arccos x$
Complementary angles	$\arccos x = \pi/2 - \arcsin x = 90° - \arcsin x$
Arccos sum	$\arccos(\alpha) + \arccos(\beta) = \arccos\left(\alpha\beta - \sqrt{(1-\alpha^2)(1-\beta^2)}\right)$
Arccos difference	$\arccos(\alpha) - \arccos(\beta) = \arccos\left(\alpha\beta + \sqrt{(1-\alpha^2)(1-\beta^2)}\right)$
Arccos of sin of x	$\arccos(\sin x) = -x - (2k + 0.5)\pi$

Sine of arccosine	$\cos(\arcsin x) = \sin(\arccos x) = \sqrt{1-x^2}$
Tangent of arccosine	$\tan(\arccos x) = \dfrac{\sqrt{1-x^2}}{x}$
Derivative of arccosine	$\dfrac{d}{dx}(\arccos x) = \arccos' x = \dfrac{-1}{\sqrt{1-x^2}}$
Indefinite integral of arccosine	$\int \arccos x\, dx = x \arccos x - \sqrt{1-x^2} + C$

Arccos Table

x	$\arccos(x)$ (rad)	$\arccos(x)$ $(°)$
−1	π	180°
$-\sqrt{3}/2$	$5\pi/6$	150°
$-\sqrt{2}/2$	$3\pi/4$	135°
−1/2	$2\pi/3$	120°
0	$\pi/2$	90°
1/2	$\pi/3$	60°
$\sqrt{2}/2$	$\pi/4$	45°
$\sqrt{3}/2$	$\pi/6$	30°
1	0	0°

Inverse Secant

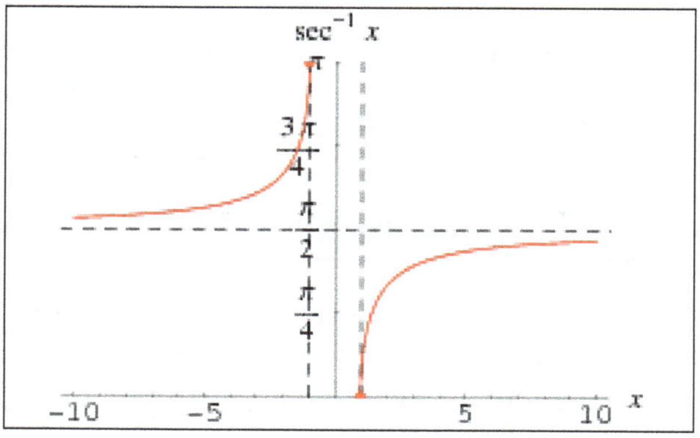

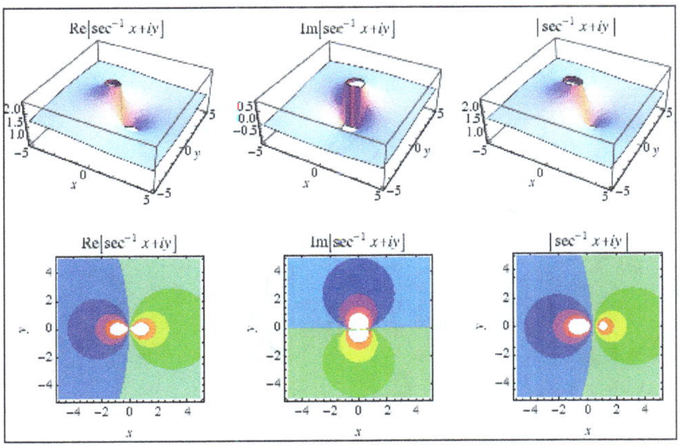

The inverse secant $\sec^{-1} z$, also denoted $\operatorname{arcsec} z$, is the inverse function of the secant. The variants $\operatorname{Arcsec} z$ and $\operatorname{Sec}^{-1} z$ are sometimes used to indicate the principal value, although this distinction is not always made. Worse yet, the notation $\operatorname{arcsec} z$ is sometimes used for the principal value, with $\operatorname{Arcsec} z$ being used for the multivalued function. In the notation $\sec^{-1} z$ (commonly used in North America and in pocket calculators worldwide), $\sec z$ is the secant and the superscript -1 denotes the inverse function, not the multiplicative inverse.

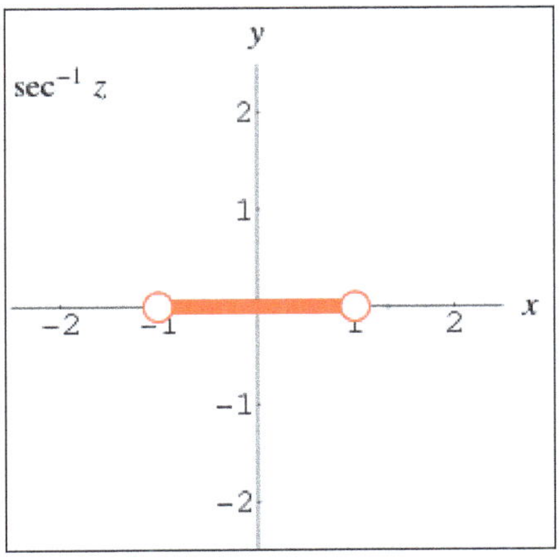

The inverse secant is a multivalued function and hence requires a branch cut in the complex plane. This follows from the definition of $\sec^{-1} z$ as:

$$\sec^{-1} z = \frac{1}{2}\pi + i \ln\left(\sqrt{1-\frac{1}{z^2}} + \frac{i}{z}\right).$$

The derivative of $\sec^{-1} z$ is:

$$\frac{d}{dz}\sec^{-1} z = \frac{1}{z^2\sqrt{1-\frac{1}{z^2}}},$$

which simplifies to:

$$\frac{d}{dx}\sec^{-1} x = \frac{1}{x\sqrt{x^2-1}}$$

for x > 0. The indefinite integral is:

$$\int \sec^{-1} z\, dz = z\sec^{-1} z - \text{In}\left[z\left(1+\sqrt{\frac{z^2-1}{z^2}}\right)\right] + C,$$

which simplifies to:

$$\int \sec^{-1} x\, dx = x\sec^{-1} x - \text{In}\left(x+\sqrt{x^2-1}\right)$$

For x > 0.

The inverse secant has a Taylor series about infinity of:

$$\begin{aligned}\sec^{-1} x &= \frac{\ }{2} \quad - \sum^{\infty} \frac{(2 \quad 1)!!}{(2n \quad 1)(2n)!!} x^{-2 \quad -1} \\ &= \frac{1}{2} \quad - x^{-1} - \frac{1}{6}x^{-3} - \frac{3}{40}x^{-5} - \frac{5}{112}x^{-7} - \ldots\end{aligned}$$

The inverse secant satisfies:

$$\sec^{-1} z = \cos^{-1}\left(\frac{1}{z}\right)$$

for $z \neq 0$, and:

$$\begin{aligned}\sec^{-1} z &= \pi - \sec^{-1}(-z) \\ &= \frac{1}{2}\pi - \csc^{-1} z \\ &= \frac{1}{2}\pi + \csc^{-1}(-z)\end{aligned}$$

for all complex z. It is given in terms of other inverse trigonometric functions by:

$$\sec^{-1} x = \begin{cases} \pi + \csc^{-1}\left(\dfrac{x}{\sqrt{x^2-1}}\right) & \text{for } x < -1 \\ \csc^{-1}\left(\dfrac{x}{\sqrt{x^2-1}}\right) & \text{for } x > 1 \end{cases}$$

$$= \begin{cases} \pi - \cot^{-1}\left(\dfrac{1}{\sqrt{x^2-1}}\right) & \text{for } x < -1 \\ \cot^{-1}\left(\dfrac{1}{\sqrt{x^2-1}}\right) & \text{for } x > 1 \end{cases}$$

$$= \begin{cases} \pi + \sin^{-1}\left(\dfrac{\sqrt{x^2-1}}{x}\right) & \text{for } x < -1 \\ \sin^{-1}\left(\dfrac{\sqrt{x^2-1}}{x}\right) & \text{for } x > 1 \end{cases}$$

$$= \begin{cases} \pi - \tan^{-1}\left(\sqrt{x^2-1}\right) & \text{for } x < -1 \\ \tan^{-1}\left(\sqrt{x^2-1}\right) & \text{for } x > 1. \end{cases}$$

Inverse Cotangent

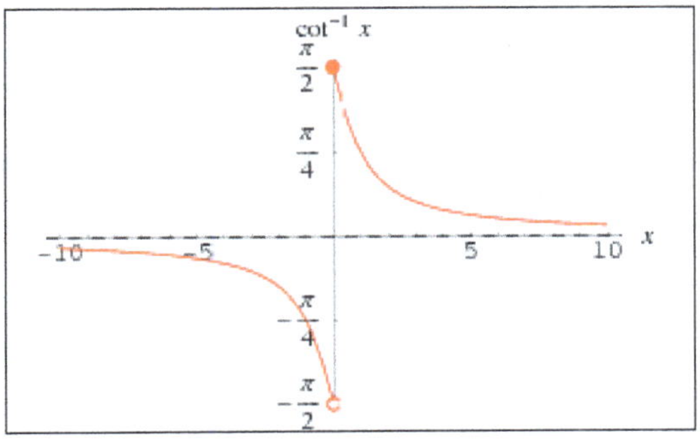

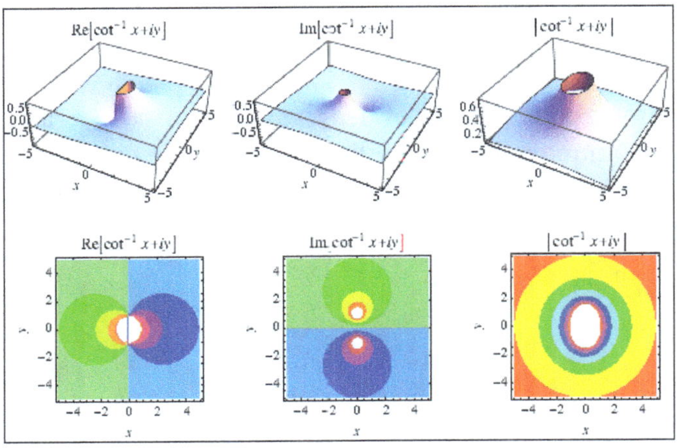

The inverse cotangent is the multivalued function $\cot^{-1} z$, also denoted $\operatorname{arccot} z$ or $\operatorname{arcctg} z$, that is the inverse function of the cotangent. The variants $\operatorname{Arc}\cot z$ and $\operatorname{Cot}^{-1} z$ are sometimes used to refer to explicit principal values of the inverse cotangent, although this distinction is not always made. Worse yet, the notation $\operatorname{arccot} z$ is sometimes used for the principal value, with $\operatorname{Arc}\cot z$ being used for the multivalued function. Note that in the notation $\cot^{-1} z$ (commonly used in North America and in pocket calculators worldwide), $\cot z$ is the cotangent and the superscript −1 denotes an inverse function, not the multiplicative inverse.

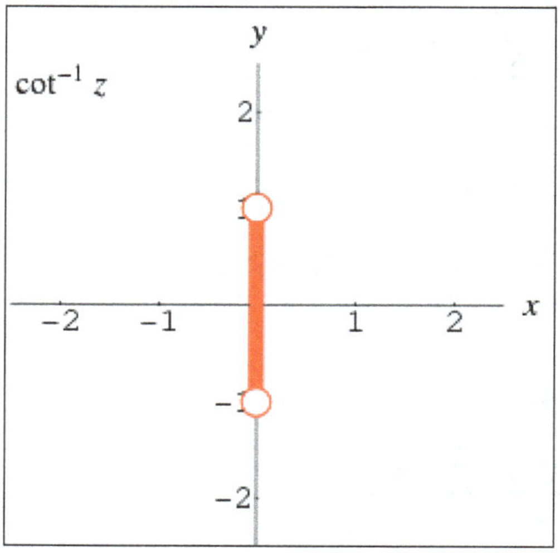

There are at least two possible conventions for defining the inverse cotangent. This work follows the convention of Abramowitz and Stegun, taking $\cot^{-1} x$ to have range $(-\pi/2, \pi/2]$, a discontinuity at $x = 0$, and the branch cut placed along the line segment $(-i, i)$. This definition can be expressed in terms of the natural logarithm by:

$$\cot^{-1} z = \frac{i}{2}\left[\ln\left(\frac{z-i}{z}\right) - \ln\left(\frac{z+i}{z}\right)\right].$$

This definition is also consistent, with the definition of ArcTan, so ArcCot[z] is equal to ArcTan[1/z].

A different but common convention defines the range of $\cot^{-1} x$ as $(0, \pi)$, thus giving a function that is continuous on the real line $\mathbb{R}$. Extreme care should be taken where examining identities involving inverse trigonometric functions, since their range of applicability or precise form may differ depending on the convention being used.

The derivative of $\cot^{-1} z$ is given by:

$$\frac{d}{dz}\cot^{-1} z = -\frac{1}{1+z^2}$$

and the integral by:

$$\int \cot^{-1} z\, d\, z = z \cot^{-1} z + \frac{1}{2} \text{In}\left(1+z^2\right) + C.$$

The Maclaurin series of the inverse cotangent for x > o is given by:

$$\cot^{-1} x = \frac{\pi}{2} - \sum_{k=0}^{\infty} \frac{(-1)^k x^{2k+1}}{2k+1}$$

$$= \frac{\pi}{2} - x + \frac{1}{3}x^3 - \frac{1}{5}x^5 + \frac{1}{7}x^7 - \frac{1}{9}x^9 + \ldots$$

The Laurent series about $z = \infty$ is given by:

$$\cot^{-1} z = \sum_{k=0}^{\infty} \frac{(-1)^k z^{-(2k+1)}}{2k+1}$$

$$= z^{-1} - \frac{1}{3}z^{-3} + \frac{1}{5}z^{-5} - \frac{1}{7}z^{-7} + \frac{1}{9}z^{-9} + \ldots$$

for $|z| > 1$.

s

Euler derived the infinite series:

$$\cot^{-1} z = \sum_{n=1}^{\infty} \frac{(2n-2)!!}{(2n-1)!!\left(z^2+1\right)^n}$$

The inverse cotangent satisfies:

$$\cot^{-1} z = \tan^{-1}\left(\frac{1}{z}\right)$$

for $z \neq 0$,

$$\cot^{-1} z = -\cot^{-1}(-z)$$

for all $z \in \mathbb{C}^*$, and:

$$\cot^{-1} x = \begin{cases} \sec^{-1}\left(\dfrac{\sqrt{x^2+1}}{x}\right) - \pi & \text{for } x > 0 \\ \sec^{-1}\left(\dfrac{\sqrt{x^2+1}}{x}\right) & \text{for } x > 0 \end{cases}$$

$$=\begin{cases} -\frac{1}{2}\pi-\tan^{-1}x & \text{for } x<0 \\ \frac{1}{2}\pi-\tan^{-1}x & \text{for } x\geq 0 \end{cases}$$

$$=\begin{cases} -\sin^{-1}\left(\frac{1}{\sqrt{x^2+1}}\right) & \text{for } x>0 \\ \sin^{-1}\left(\frac{1}{\sqrt{x^2+1}}\right) & \text{for } x>0 \end{cases}$$

$$=\begin{cases} -\frac{1}{2}\pi-\cot^{-1}\left(\frac{1}{x}\right) & \text{for } x>0 \\ \frac{1}{2}\pi-\cot^{-1}\left(\frac{1}{x}\right) & \text{for } x>0 \end{cases}$$

$$=\begin{cases} -\csc^{-1}\left(\sqrt{x^2+1}\right) & \text{for } x>0 \\ \csc^{-1}\left(\sqrt{x^2+1}\right) & \text{for } x>0 \end{cases}$$

$$=\begin{cases} \cos^{-1}\left(\frac{x}{\sqrt{x^2+1}}\right)-\pi & \text{for } x>0 \\ \cos^{-1}\left(\frac{x}{\sqrt{x^2+1}}\right) & \text{for } x>0 \end{cases}$$

$$=\begin{cases} -\frac{1}{2}\pi-\sin^{-1}\left(\frac{x}{\sqrt{x^2+1}}\right) & \text{for } x<0 \\ \frac{1}{2}\pi-\sin^{-1}\left(\frac{x}{\sqrt{x^2+1}}\right) & \text{for } x>0 \end{cases}$$

Analytic sums of cotangents include the beautiful result:

$$\sum_{n=1}^{\infty}\cot^{-1}n^2=\cot^{-1}\left(\frac{1+t}{1-t}\right)=1.42474...,$$

Where:

$$t\equiv\cot\left(\frac{1}{2}\pi\sqrt{2}\right)\tanh\left(\frac{1}{2}\pi\sqrt{2}\right)$$

A number:

$$t_x = \cot^{-1} x,$$

where x is an integer or rational number, is sometimes called a Gregory number. Lehmer showed that $\cot^{-1}(a/b)$ can be expressed as a finite sum of inverse cotangents of integer arguments:

$$\cot^{-1}\left(\frac{a}{b}\right) = \sum_{i=1}^{k} (-1)^{i=1} \cot^{-1} n_i,$$

Where:

$$n_i = \left\lfloor \frac{a_i}{b_i} \right\rfloor,$$

with $\lfloor x \rfloor$ the floor function, and:

$$a_{i+1} = a_i n + i + b_i$$
$$b_{i+1} = a_i - n_i b_i,$$

with $a_0 = a$ and $b_0 = b$, and where the recurrence is continued until $b_{k+1} = 0$. If an inverse tangent sum is written as:

$$\tan^{-1} n = \sum_{k=1} f_k \tan^{-1} n_k + f \tan^{-1} 1,$$

then equation becomes:

$$\cot^{-1} n = \sum_{k=1} f_k \cot^{-1} n_k + c \cot^{-1} 1,$$

Where:

$$c = 2 - f - 2\sum_{k=1} f_k.$$

Inverse cotangent sums can be used to generate Machin-like formulas.

Other inverse cotangent identities include:

$$2\cot^{-1}(2x) - \cot^{-1} x = \cot^{-1}(4x^3 + 3x)$$
$$2\cot^{-1}(3x) - \cot^{-1} x = \cot^{-1}\left(\frac{27x^4 + 18x^2 - 1}{8x}\right)$$

as well as many others. Note that for equation $2\cot^{-1}(3x)-\cot^{-1}x=\cot^{-1}\left(\frac{27x^4+18x^2-1}{8x}\right)$, the choice of convention for $\cot^{-1}z$ is significant, since it holds for all complex z in the $[0,\pi]$ convention, but holds only outside a lens-shaped region centered on the origin in the $[-\pi/2,\pi/2]$ convention.

Properties of Inverse Trigonometric Functions

Property 1:

- $\sin^{-1}(1/x)=\text{cosec}^{-1}x,\ x\geq 1 \text{ or } x\leq -1$
- $\cos^{-1}(1/x)=\sec^{-1}x,\ x\geq 1 \text{ or } x\leq -1$
- $\tan^{-1}(1/x)=\cot^{-1}x,\ x>0$

Proof:

$\sin^{-1}(1/x)=\text{cosec}^{-1}x,\ x\geq 1 \text{ or } x\leq -1$, Let $\sin-1x=y$

i.e. x = cosec y

$$\frac{1}{x}=\sin y$$

$$\sin^{-1}\frac{1}{x}\Big)=y$$

$$\sin^{-1}\frac{1}{x}\Big)=\text{cosec}^{-1}x$$

$$\sin^{-1}\left(\frac{1}{x}\right)=\text{cosec}^{-1}x$$

Hence, $\sin^{-1}\frac{1}{x}=\text{cosec}^{-1}x$ where, x ≥ 1 or x ≤ -1.

Property 2:

- $\sin^{-1}(-x)=-\sin^{-1}(x),\ x\in[-1,1]$
- $\tan^{-1}(-x)=-\tan^{-1}(x),\ x\in R$
- $\text{cosec}^{-1}(-x)=-\text{cosec}^{-1}(x),\ |x|\geq 1$

Proof:

$$\sin^{-1}(-x) = -\sin^{-1}(x),\ \ x \in [-1,1]$$ Let, $\sin{-1}(-x) = y$

Then $-x = \sin y$

$$x = -\sin y$$
$$x = \sin(-y)$$
$$\sin^{-1} = \sin^{-1}(\sin(-y))$$
$$\sin^{-1} x = y$$
$$\sin^{-1} x = -\sin^{-1}(-x)$$

Hence, $\sin^{-1}(-x) = -\sin^{-1} x \in [-1,1]$

Property 3:

- $\cos^{-1}(-x) = \pi - \cos^{-1} x,\ x \in [-1,1]$
- $\sec^{-1}(-x) = \pi - \sec^{-1} x,\ \ |x| \geq 1$
- $\cot^{-1}(-x) = \pi - \cot^{-1} x,\ x \in R$

Proof:

$$\cos^{-1}(-x) = \pi - \cos^{-1} x,\ x \in [-1,1]$$

Let $\cos^{-1}(-x) = y$

$$\cos y = -x \ \ x = -\cos y$$
$$x = \cos(\pi - y)$$

Since, $\cos \pi - q = -\cos q$

$$\cos^{-1} x = \pi - y$$
$$\cos^{-1} x = \pi - \cos^{-1} - x$$

Hence, $\cos^{-1} - x = \pi - \cos^{-1} x$

Property 4:

- $\sin^{-1} x + \cos^{-1} x = \pi / 2,\ x \in [-1,1]$
- $\tan^{-1} x + \cot^{-1} x = \pi / 2,\ x \in R$
- $\cos ec^{-1} \mathrm{x} + \sec^{-1} \mathrm{x} = \pi / 2,\ \ |\mathrm{x}| \geq 1$

Proof:

$$\sin^{-1}x+\cos^{-1}x=\pi/2,\ x\in[-1,1]$$

Let $\sin-1x = y$ or $x = \sin y = \cos\left(\frac{\pi}{2}-y\right)$

$$\cos^{-1}x=\cos^{-1}\left(\cos\left(\frac{\pi}{2}-y\right)\right)$$

$$\cos^{-1}x=\frac{\pi}{2}-y$$

$$\cos^{-1}x=\frac{\pi}{2}-\sin^{-1}x$$

$$\sin^{-1}+\cos^{-1}x=\frac{\pi}{2}$$

Hence, $\sin^{-1}x+\cos^{-1}x=\pi/2,\ x\in[-1,1]$

Property 5:

- $\tan^{-1}x+\tan^{-1}y=\tan^{-1}((x+y)/(1-xy)),\ xy<1.$
- $\tan^{-1}x-\tan^{-1}y=\tan^{-1}((x-y)/(1+xy)),\ xy>-1.$

Proof:

$$\tan^{-1}x+\tan^{-1}y=\tan^{-1}((x+y)/(1-xy)),\ xy<1.$$

Let $\tan^{-1}x=A$

And $\tan^{-1}y=B$

Then, $\tan A=x$

$\tan B=y$

Now, $\tan(A+B)=(\tan A+\tan B)/(1-\tan A\tan B)$

$$\tan(A+B)=\frac{x+y}{1-xy}$$

$$\tan^{-1}\left(\frac{x+y}{1-xy}\right)=A+B$$

Hence, $\tan^{-1}\left(\frac{x+y}{1-xy}\right)=\tan^{-1}x+\tan^{-1}y$

Property 6:

- $2\tan^{-1} x = \sin^{-1}\left(2x/\left(1+x^2\right)\right),\ |x| \le 1$
- $2\tan^{-1} x = \cos^{-1}\left(\left(1-x^2\right)/\left(1+x^2\right)\right),\ x \ge 0$
- $2\tan^{-1} x = \tan^{-1}\left(2x/\left(1-x^2\right)\right), -1 < x < 1$

Proof:

$$2\tan^{-1} x = \sin^{-1}\left(2x/\left(1+x^2\right)\right),\ |x| \le 1$$

Let $\tan^{-1} x = y$ and $x = \tan y$

Consider RHS. $\sin^{-1}\left(\dfrac{2x}{1+x^2}\right)$

$$= \sin^{-1}\left(\frac{2\tan y}{1+\tan^2 y}\right)$$

$$= \sin^{-1}\left(\sin 2y\right)$$

Since, $\sin 2\theta = 2\tan\theta/\left(1+\tan^2\theta\right)$,

= 2y $= 2\tan^{-1} x$ which is our LHS

Hence $2\tan^{-1} x = \sin^{-1}\left(2x/\left(1+x^2\right)\right),\ |x| \le 1$

Solved Example:

Prove that "$\sin^{-1}(-x) = -\sin^{-1}(x),\ x \in [-1,1]$"

Ans: Let, $\sin^{-1}(-x) = y$

Then $-x = \sin y$

$$x = -\sin y$$

$$x = \sin(-y)$$

$$\sin^{-1} x = \arcsin\left(\sin(-y)\right)$$

$$\sin^{-1} x = y$$

$$\sin^{-1} x = -\sin^{-1}(-x)$$

Hence, $\sin^{-1}(-x) = -\sin^{-1} x,\ x \in [-1,1]$

References

- InverseSine: mathworld.wolfram.com, Retrieved 13 June, 2019
- Arcsin, math-trigonometry: rapidtables.com, Retrieved 14 March, 2019
- Properties-inverse-functions, maths-inverse-trigonometric-functions: toppr.com, Retrieved 28 April, 2019
- InverseCotangent: mathworld.wolfram.com, Retrieved 24 July, 2019
- Arccos, math-trigonometry: rapidtables.com, Retrieved 31 August, 2019
- InverseSecant: mathworld.wolfram.com, Retrieved 13 May, 2019

4

Graphs of Trigonometric Functions

The graphs of trigonometric functions of sine, cosine, tangent, cotangent, secant, and cosecant functions are periodic in nature. The graph cycle is repeated after every angle of 180 degrees. This chapter closely examines these graphs of trigonometric functions to provide an extensive understanding of the subject.

Graph of Sin x

y = sin x is periodic function. The period of y = sin x is 2π. Therefore, we will draw the graph of y = sin x in the interval $[-\pi, 2\pi]$.

For this, we need to take the different values of x at intervals of 10°. Then by using the table of natural sines we will get the corresponding values of sin x. Take the values of sin x correct to two place of decimal. The values of sin x for the different values of x in the interval $[-\pi, 2\pi]$ are given in the following table.

We draw two mutually perpendicular straight lines XOX' and YOY'. XOX' is called the x-axis which is a horizontal line. YOY' is called the y-axis which is a vertical line. Point O is called the origin.

Now represent angle (x) along x-axis and y (or sin x) along y-axis.

Along the x-axis: Take 1 small square = 10^c.

Along the y-axis: Take 10 small squares = 1 unity.

Now plot the above tabulated values of x and y on the co-ordinate graph paper. Then join the points by free hand. The continuous curve obtained by free hand joining is the required graph of y = sin x.

Steps to Draw the Graph of y = c Sin ax

Step I: Obtain the values of a and c.

Step II: Draw the graph of y = sin x and mark the points where y = sin x crosses x-axis.

Step III: Divide the x-coordinate of the points where y = sin x crosses x-axis by a and mark maximum and minimum values of y = c sin ax as c and –c on y-axis.

The graph obtained is the required graph of y = c sin ax.

Properties of y = Sin x

(i) The graph of the function y = sin x is continuous and extends on either side in symmetrical wave form.

(ii) Since the graph intersects the x-axis at the origin and at points where x is an even multiple of 90°, hence sin x is zero at x = nπ where n = 0, ±1, ±2, ±3, ±4, ...

(iii) The ordinate of any point on the graph always lies between 1 and – 1 i.e., $-1 \le y \le 1$ or ,$-1 \le \sin x \le 1$ hence, the maximum value of sin x is 1 and its minimum value is – 1 and these values occur alternately at $\frac{\pi}{2}, \frac{3\pi}{2}, \frac{5\pi}{2}$,.... i. e., at $x = (2n+1)\frac{\pi}{2}$, where n = 0, ±1, ±2, ±3, ±4, ...

(iv) Since the function y= sin x is periodic of period 2π, hence the portion of the graph between 0 to 2π is repeated over and over again on either side.

The following table shows the value of sin x for various values of x. (Namely all multiples of 30 degrees and 45 degrees, except we're using radians).

x	0	$\frac{\pi}{6}$	$\frac{\pi}{4}$	$\frac{\pi}{3}$	$\frac{\pi}{2}$	$\frac{2\pi}{3}$	$\frac{3\pi}{4}$	$\frac{5\pi}{6}$	π	$\frac{7\pi}{6}$	$\frac{5\pi}{4}$	$\frac{4\pi}{3}$	$\frac{3\pi}{2}$	$\frac{5\pi}{3}$	$\frac{7\pi}{4}$	$\frac{11\pi}{4}$	2π
y=sin(x)	0	$\frac{1}{2}$	$\frac{1}{\sqrt{2}}$	$\frac{\sqrt{3}}{2}$	1	$\frac{\sqrt{3}}{2}$	$\frac{1}{\sqrt{2}}$	$\frac{1}{2}$	0	$-\frac{1}{2}$	$-\frac{1}{\sqrt{2}}$	$-\frac{\sqrt{3}}{2}$	-1	$\frac{\sqrt{3}}{2}$	$-\frac{1}{\sqrt{2}}$	$-\frac{1}{2}$	0

If we plot these points (x, y) they look like this:

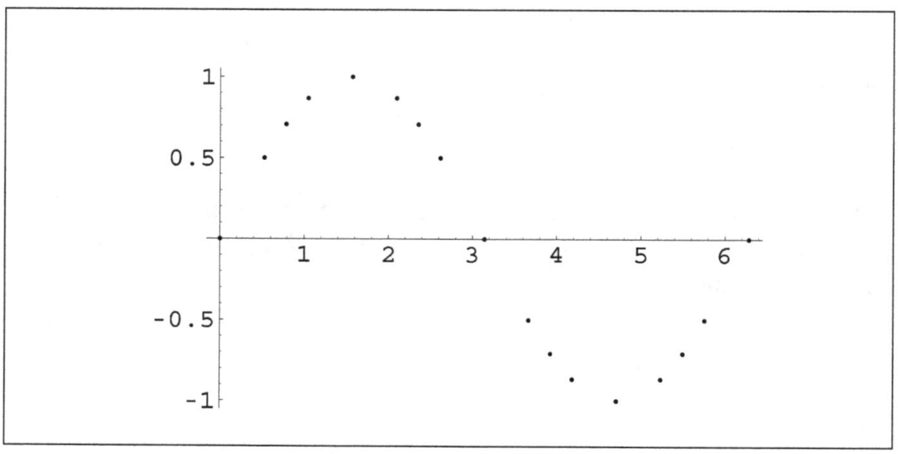

If we connect the dots using a smooth curve, we'll get the following graph.

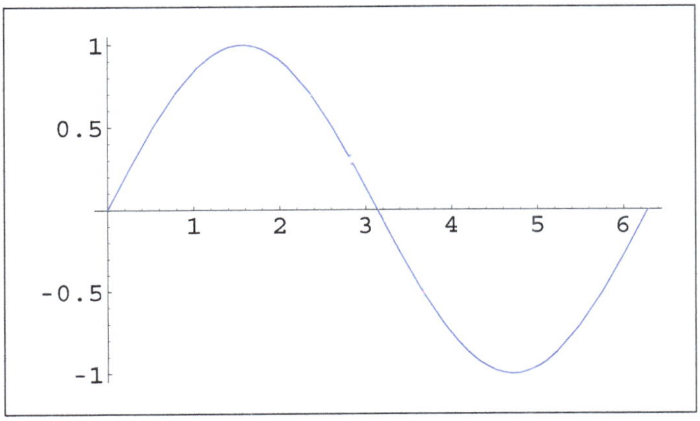

We know that sin (x) is periodic with period 2π. That means the graph just repeats forever and ever to the left and right.

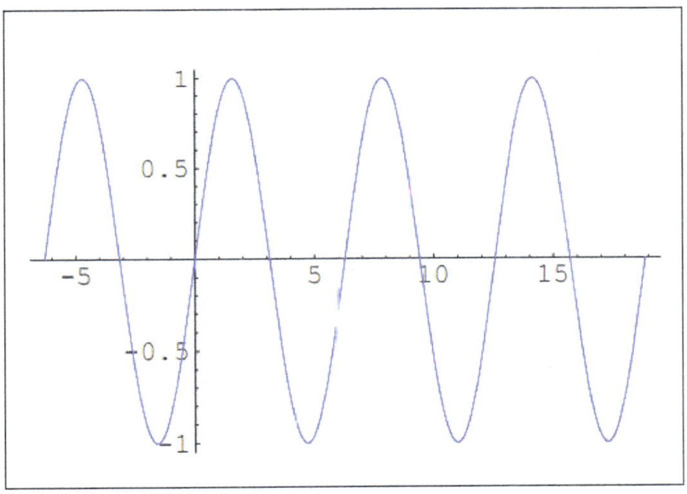

Graph of Cos x

y = cos x is periodic function. The period of y = cos x is 2π. Therefore, we will draw the graph of y = cos x in the interval [−π, 2π].

For this, we need to take the different values of x at intervals of 10°. Then by using the table of natural cosines we will get the corresponding values of cos x. Take the values of cos x correct to two place of decimal. The values of cos x for the different values of x in the interval [−π, 2π] are given in the following table.

We draw two mutually perpendicular straight lines XOX' and YOY'. XOX' is called the x-axis which is a horizontal line. YOY' is called the y-axis which is a vertical line. Point O is called the origin.

Now represent angle (x) along x-axis and y (or cos x) along y-axis.

Along the x-axis: Take 1 small square = 10°.

Along the y-axis: Take 10 small squares = 1 unity.

Now plot the above tabulated values of x and y on the co-ordinate graph paper. Then join the points by free hand. The continuous curve obtained by free hand joining is the required graph of y = cos x.

Steps to Draw the Graph of y = c Cos ax

Steps I: Obtain the values of a and c.

Step II: Draw the graph of y = cos x and mark the points where y = cos x crosses x-axis.

Step III: Divide the x-coordinate of the points where y = cos x crosses x-axis by a and mark maximum and minimum values of y = c cos ax as c and –c on y-axis.

The graph obtained is the required graph of y = c cos ax.

Properties of y = Cos x

(i) The graph of the function y = cos x is continuous and extends on either side in symmetrical wave form.

(ii) Since the graph of y = cos x intersects the x-axis at the origin and at points where x is an odd multiple of 90°, hence cos x is zero at $x = (2n + 1)\frac{\pi}{2}$ where n = 0, ±1, ±2, ±3, ±4, ...

(iii) The ordinate of any point on the graph always lies between 1 and - 1 i.e., $-1 \le y \le 1$ or, $-1 \le \cos x \le 1$ hence, the maximum value of cos x is 1 and its minimum value is - 1 and these values occur alternately at x = 0, π, 2π,... i. e., at x = nπ, where n = 0, ±1, ±2, ±3, ±4, ...

(iv) The portion of the graph between 0 to 2π is repeated over and over again on either side, since the function y = cos x is periodic of period 2π.

The following table shows the value of cos x for various values of x. (Namely all multiples of 30 degrees and 45 degrees, except we're using radians.)

x	0	$\frac{\pi}{6}$	$\frac{\pi}{4}$	$\frac{\pi}{3}$	$\frac{\pi}{2}$	$\frac{2\pi}{3}$	$\frac{3\pi}{4}$	$\frac{5\pi}{6}$	π	$\frac{7\pi}{6}$	$\frac{5\pi}{4}$	$\frac{4\pi}{3}$	$\frac{3\pi}{2}$	$\frac{5\pi}{3}$	$\frac{7\pi}{4}$	$\frac{11\pi}{6}$	2π
y-cos(x)	1	$\frac{\sqrt{3}}{2}$	$\frac{1}{\sqrt{2}}$	$\frac{1}{2}$	0	$-\frac{1}{2}$	$-\frac{1}{\sqrt{2}}$	$-\frac{\sqrt{3}}{2}$	-1	$-\frac{\sqrt{3}}{2}$	$-\frac{1}{\sqrt{2}}$	$-\frac{1}{2}$	0	$\frac{1}{2}$	$\frac{1}{\sqrt{2}}$	$\frac{\sqrt{3}}{2}$	1

If we plot these points (x, y) they look like this:

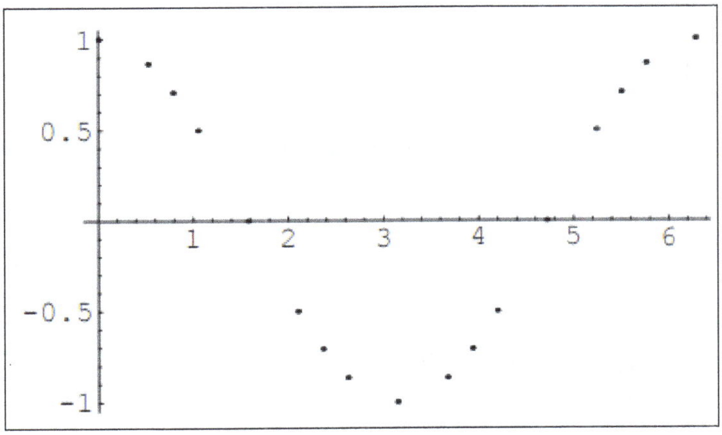

If we connect the dots using a smooth curve, we'll get the following graph.

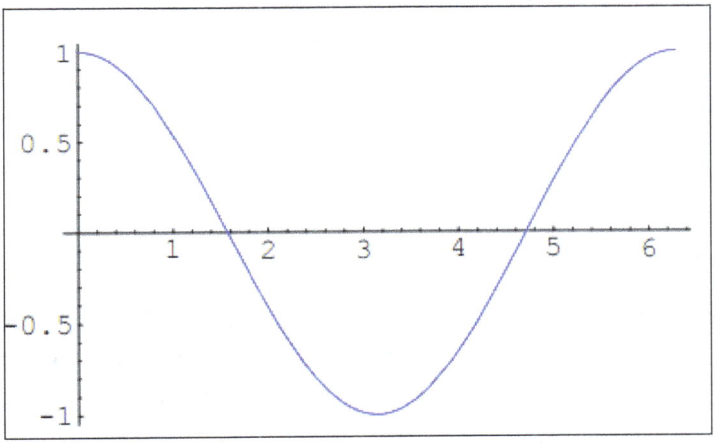

We know that cos (x) is periodic with period 2π. That means the graph just repeats forever and ever to the left and right.

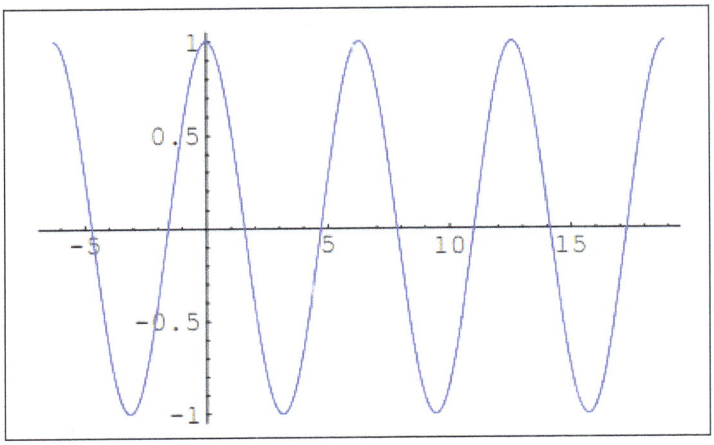

Graph of Tan x

The tangent function has a parent graph just like any other function. Using the graph of this function, you can make the same type of transformation that applies to the parent graph of any function. The easiest way to remember how to graph the tangent function is to remember that:

$\tan\theta = \dfrac{\sin\theta}{\cos\theta}$. Because $\cos\theta = 0$ for various values of θ,

some interesting things happen to tangent's graph. When the denominator of a fraction is 0, the fraction is *undefined.* Therefore, the graph of tangent has asymptotes, which is where the function is undefined, at each of these places.

The table presents $\theta, \sin\theta$, and $\tan\theta$.

It shows the roots (or zeros), the asymptotes (where the function is undefined), and the behavior of the graph in between certain key points on the unit circle.

θ	0	$0<\theta<\frac{\pi}{2}$	$\frac{\pi}{2}$	$\frac{\pi}{2}<\theta<\pi$	π	$\pi<\theta<\frac{3\pi}{2}$	$\frac{3\pi}{2}$	$\frac{3\pi}{2}<\theta<2\pi$	2π
$\sin\theta$	0	positive	1	positive	0	negative	-1	negative	0
$\cos\theta$	1	positive	0	negative	-1	negative	0	positive	1
$\tan\theta$	0	positive	undef.	negative	0	positive	undef.	negative	0

To plot the parent graph of a tangent function $f(x) = \tan x$ where x represents the angle in radians, you start out by finding the vertical asymptotes. Those asymptotes give you some structure from which you can fill in the missing points.

- Find the vertical asymptotes so you can find the domain.

These steps use x instead of theta because the graph is on the x–y plane. In order to find the domain of the tangent function $f(x) = \tan x$, you have to locate the vertical asymptotes. The first asymptote occurs when the angle:

$x = \dfrac{\pi}{2}$, and it repeats every π radians.

The period of the tangent graph is:

π radians.

which is different from that of sine and cosine.) Tangent, in other words, has asymptotes when:

$X = \dfrac{\pi}{2}$ and $\dfrac{3\pi}{2}$.

The easiest way to write this is:

$$X \neq \frac{\pi}{2} + n\pi$$

where n is an integer. You write:

$"+n\pi"$ because the period of tangent is π radians,

so if an asymptote is at:

$\frac{\pi}{2}$ and you add or subtract π,

you automatically find the next asymptote.

- Determine values for the range.

Recall that the tangent function can be defined as:

$$\frac{\sin x}{\cos x}.$$

The closer you get to the values where:

$$\cos x = 0,$$

the smaller the number on the bottom of the fraction gets and the larger the value of the overall fraction gets — in either the positive or negative direction.

The range of tangent has no restrictions; you aren't stuck between 1 and −1, like with sine and cosine. In fact, the ratios are any and all numbers. The range is:

$$(-\infty, \infty).$$

- Calculate the graph's x-intercepts.

Tangent's parent graph has roots (it crosses the x-axis) at:

$$0, \pi, \text{ and } 2\pi.$$

You can find these values by setting:

$$\frac{\sin x}{\cos x}$$

equal to 0 and then solving. The x-intercepts for the parent graph of tangent are located wherever the sine value is 0.

- Figure out what's happening to the graph between the intercepts and the asymptotes.
    - The graph of $f(x) = \tan x$ is positive for angles in the first quadrant (with respect to the unit circle) and points upward toward the asymptote at pi/2, because all sine and cosine values are positive for angles in the first quadrant.
    - The graph of $f(x) = \tan x$ is negative for angles in Quadrant II because sine is positive and cosine is negative for angles in this quadrant.
    - The graph of $f(x) = \tan x$ is positive for angles in Quadrant III because both sine and cosine are negative.
    - Finally, the graph of $f(x) = \tan x$ is positive for angles in Quadrant IV because sine is negative and cosine is positive for angles in this quadrant.

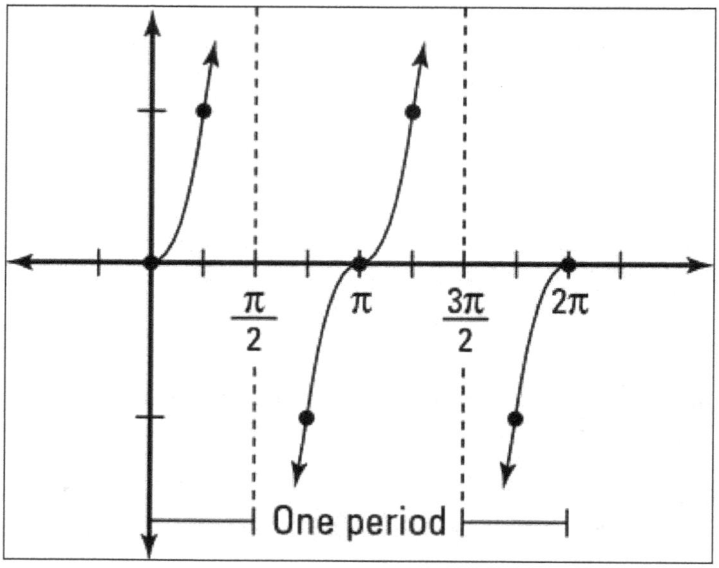

A tangent graph has no maximum or minimum points.

$$f(x) = \tan x.$$

Graph of Cosec x

y = csc x is periodic function. The period of y = csc x is 2π. Therefore, we will draw the graph of y = csc x in the interval [-π, 2π].

For this, we need to take the different values of x at intervals of 10°. Then by using the table of natural sines we will get the corresponding values of csc x. Take the values of sin x correct to two place of decimal. The values of csc x for the different values of x in the interval [-π, 2π] are given in the following table.

We draw two mutually perpendicular straight lines XOX' and YOY'. XOX' is called the x-axis which is a horizontal line. YOY' is called the y-axis which is a vertical line. Point O is called the origin.

Now represent angle (x) along x-axis and y (or csc x) along y-axis.

Along the x-axis: Take 1 small square = 10°.

Along the y-axis: Take 10 small squares = 1 unity.

Now plot the above tabulated values of x and y on the co-ordinate graph paper. Then join the points by free hand. The continuous curve obtained by free hand joining is the required graph of y = csc x.

The graph of the cosecant function appears as:

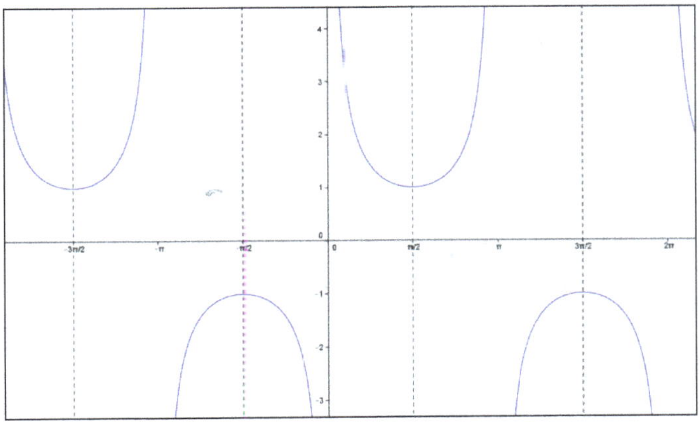

Properties of y = Csc x

- The graph of the function y = csc x is not a continuous graph, but consists of infinite number of separate branches, the points of discontinuities are at x = nπ, where n = 0, ±1, ±2, ±3, ±4, ...
- As x passes through any point of discontinuity from the left to the right, the value of csc x suddenly changes from (-∞) to (+ ∞).
- Each branch of the curve approaches continuously the two lines parallel to y-axis at two points of discontinuity of the graph. Such lines are called asymptotes to the curve.
- No part of the graph lies between the lines y = 1 and y = -1, since |csc x| ≥ 1.
- The portion of the graph between 0 to 2π is repeated over and over again on either side, since the function y = csc x is periodic of period 2π.

Graph of Sec x

y = sec x is periodic function. The period of y = sec x is 2π. Therefore, we will draw the graph of y = sec x in the interval [-π, 2π].

For this, we need to take the different values of x at intervals of 10°. Then by using the table of natural cosines we will get the corresponding values of cos x. Take the values of cos x correct to two place of decimal. The values of cos x for the different values of x in the interval [-π, 2π] are given in the following table.

We draw two mutually perpendicular straight lines XOX' and YOY'. XOX' is called the x-axis which is a horizontal line. YOY' is called the y-axis which is a vertical line. Point O is called the origin.

Now represent angle (x) along x-axis and y (or sec x) along y-axis.

Along the x-axis: Take 1 small square = 10°.

Along the y-axis: Take 10 small squares = 1 unity.

Now plot the above tabulated values of x and y on the co-ordinate graph paper. Then join the points by free hand. The continuous curve obtained by free hand joining is the required graph of y = sec x.

The graph of the secant function appears as:

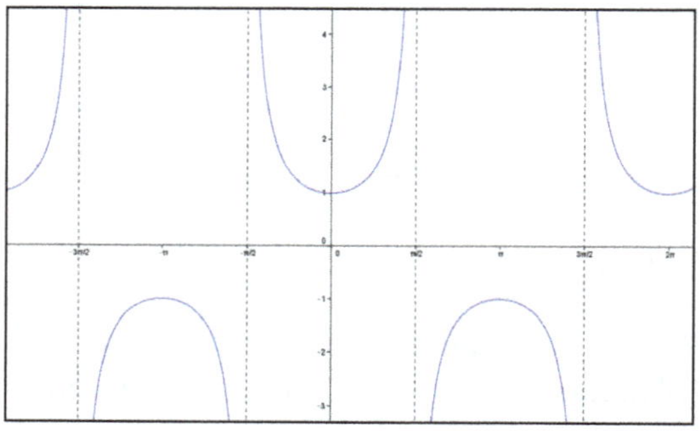

Properties of y = Sec x

- The graph of the function y = cos x is not a continuous graph, but consists of infinite number of separate branches, the points of discontinuities are at x = (2n + 1)π2π2, where n = 0, ±1, ±2, ±3, ±4, ...

  The straight lines parallel to y-axis at these points of discontinuities are asymptotes to the different branches of the curve.

- Comparing cosecant-graph and secant-graph we see that cosecant-graph coincides with secant-graph if the former is shifted to the left through 90° this is due to the fact that cos (90° + x) = sec x.

- No part of the graph lies between the lines = 1 and y = -1, since |sec x| ≥ 1.
- The portion of the graph between 0 to 2π is repeated over and over again on either side, since the function y = sec x is periodic of period 2π.

Graph of Cot x

y = cot x is periodic function. The period of y = cot x is π. Therefore, we will draw the graph of y = cot x in the interval [-π, 2π].

For this, we need to take the different values of x at intervals of 10°. Then by using the table of natural cotangent we will get the corresponding values of cot x. Take the values of cot x correct to two place of decimal. The values of cot x for the different values of x in the interval [-π, 2π] are given in the following table.

We draw two mutually perpendicular straight lines XOX' and YOY'. XOX' is called the x-axis which is a horizontal line. YOY' is called the y-axis which is a vertical line. Point O is called the origin.

Now represent angle (x) along x-axis and y (or tan x) along y-axis.

Along the x-axis: Take 1 small square = 10°.

Along the y-axis: Take 10 small squares = 1 unity.

Now plot the above tabulated values of x and y on the co-ordinate graph paper. Then join the points by free hand. The continuous curve obtained by free hand joining is the required graph of y = cot x.

The graph of the cotangent function appears as:

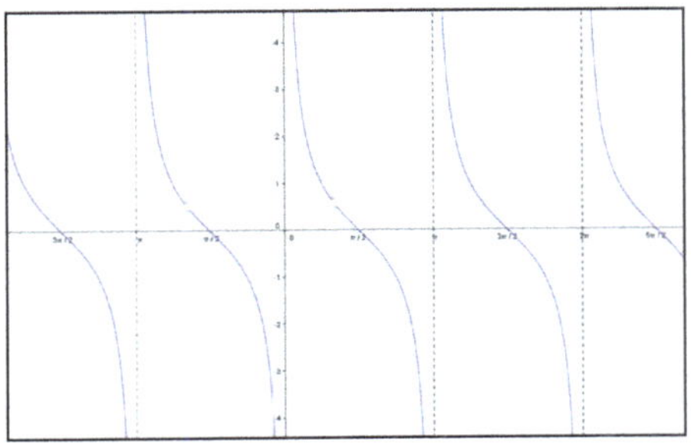

Properties of y = Cot x

- The cotangent-graph is not a continuous graph, but consist of infinite separate branches parallel to one another, the points of discontinuities are at $x = n\pi$, where $n = 0, \pm1, \pm2, \pm3, \pm4, \ldots$

- As x passes through any point of discontinuities from the late to the right, the value of cot x suddenly changes from $(-\infty)$ to $(+\infty)$.
- Each branch of the curve approaches continuously the two lines are called asymptotes to the curve.
- Each branch is simply a repetition of the branch from 0° to 180°, Since the function y= cot x is periodic of period π.

Graphing Trigonometric Functions

You've already learned the basic trigonometric graphs. But just as you could make the basic quadratic, $y=x^2$, more complicated, such as $y= -(x+5)^2 - 3$, so also trigonometric graphs can be made more complicated. We can transform and translate trigonometric functions, just like you transformed and translated other functions in algebra.

Let's start with the basic sine function, $f(t) = sin(t)$. This function has an amplitude of 1 because the graph goes one unit up and one unit down from the midline of the graph. This function has a period of 2πbecause the sine wave repeats every 2π units.

The graph looks like this:

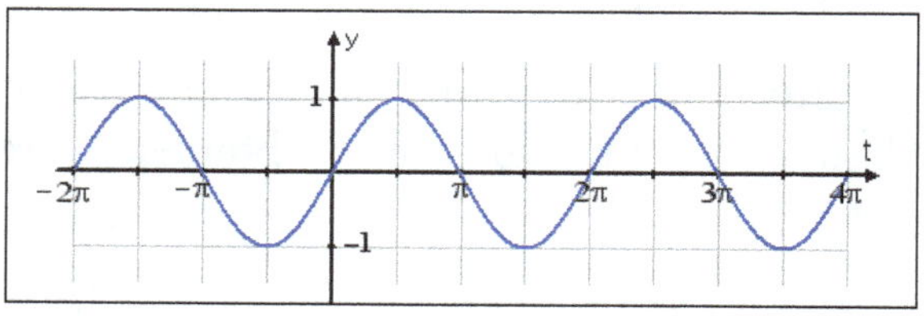

Now let's look at $g(t) = 3sin(t)$:

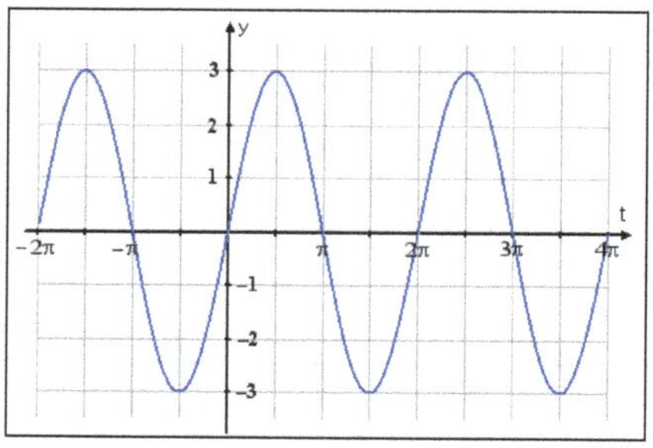

Do you see that this second graph is three times as tall as was the first graph? The amplitude has changed from 1 in the first graph to 3 in the second, just as the multiplier in front of the sine changed from 1 to 3. This relationship is always true: Whatever number A is multiplied on the trigonometric function gives you the amplitude (that is, the "tallness" or "shortness" of the graph); in this case, that amplitude number was 3.

- What is the amplitude of $y(t) = 0.5cos(t)$?

For this function, the value of the amplitude multiplier A is given by 0.5, so the function will have an amplitude of:

0.5 = 1/2

- What is the amplitude of y(x) = −2cos(x)?

For this function, the value of the amplitude multiplier A is −2, so the amplitude is:

2

and, by the way, the graph would also be flipped upside down, because of the "minus" sign.

Technically, the amplitude is the absolute value of whatever is multiplied on the trigonometric function. The amplitude just says how "tall" or "short" the curve is; it's up to you to notice whether there's a "minus" on that multiplier, and thus whether or not the function is in the usual orientation, or upside-down.

Recall the first graph, being the "regular" sine wave:

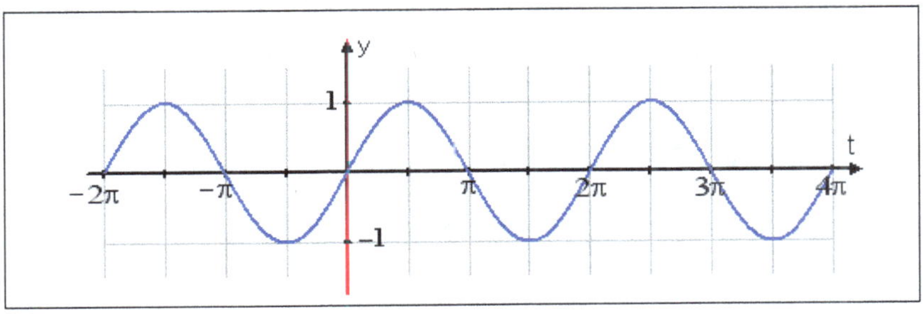

Now let's look at $h(t) = sin(2t)$:

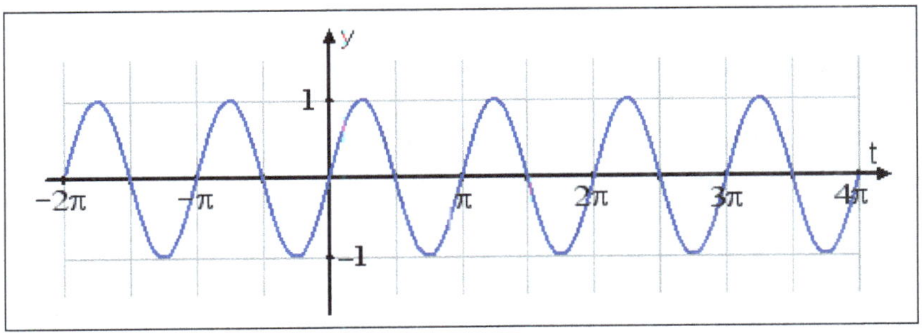

Do you see how this third graph is squished in from the sides, as compared with the first graph? Do you see that the sine wave is cycling twice as fast, so its period is only half as long? This relationship is always true: Whatever value B is multiplied on the variable (inside the trigonometric function), you use this value to find the period ω ("omega", not "double-u") of the trigonometric function, according to this formula:

- General period formula:

$$\omega = \frac{(\text{regular period})}{|B|}$$

For sines and cosines (and their reciprocals), the "regular" period is 2π, so their formula is:

- Period formula for sines & cosines:

$$\omega = \frac{2\pi}{|B|}$$

For tangents and cotangents, the "regular" period is π, so their formula is:

- Period formula for tangents & cotangents:

$$\omega = \frac{\pi}{|B|}$$

In the sine wave graphed above, the value of the period multiplier B was 2. (Sometimes the value of B inside the function will be negative, which is why there are absolute-value bars on the denominator.) As a result, its period was 2π/2 = π.

- What is the period of $f(t) = cos(3t)$?

The formula for sines and cosines says that the regular period is 2π. In $cos(3t)$, B = 3, so this function will have a period of:

$$\frac{2\pi}{3} = \left(\frac{2}{3}\right)\pi$$

- What is the period of $g(x) = tan(x/2)$?

The formula for tangents and cotangents says that the regular period is π. For $tan(t/2)$, the value of B is 1/2, so the function will have a period of:

$$\frac{\pi}{\left(\frac{1}{2}\right)} = 2\pi$$

Recall again the first graph, being the "regular" sine wave:

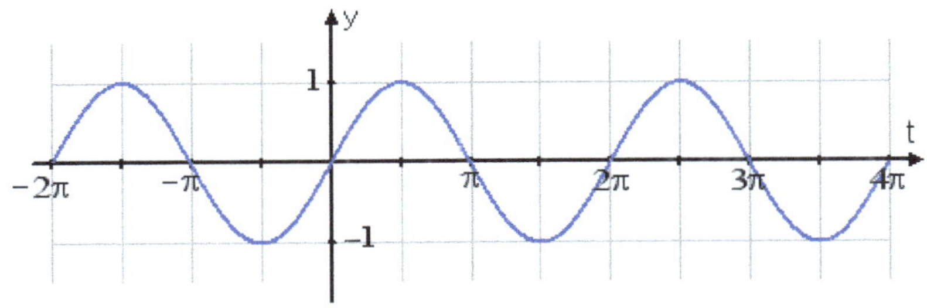

Now let's looks at *j(t)* = *sin*(*t* – π/3):

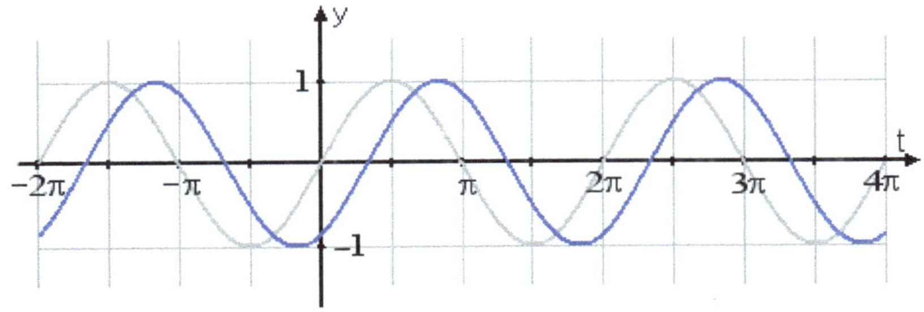

Do you see that the graph (shown in blue on the graph above) is shifted over to the right by π/3 units from the regular graph (shown in gray)? This relationship is always true: If the argument of the function (the thing you're plugging in to the function) is of the form "(variable) – (number) = (variable) – C", then the graph is shifted to the *right* by that (number) of units (that is, by C units); if the argument is of the form "(variable) + (number) = (variable) + C", then the graph is shifted to the *left* by that (number) of units (again, by C units). This right- or left-shifting is called "phase shift".

- What is the phase shift of $y(t) = \cos\left(x + \frac{\pi}{4}\right)$?

Inside the argument (that is, inside the parentheses of the function), a π/4 is added to the variable. This means that C = π/4. Because this value is *added* to the variable, then the shift is to the *left*. Then the phase shift is:

to the left by $\frac{\pi}{4}$

- What is the phase shift of $f(t) = \tan\left(t - \frac{2\pi}{3}\right)$

The number C inside with the variable is 2π/3, so this will be the phase shift. This number is *subtracted* from the variable, so the shift will be to the *right*.

to the right by $\frac{2\pi}{3}$

Let's recollect again the graph of the "regular" sine wave:

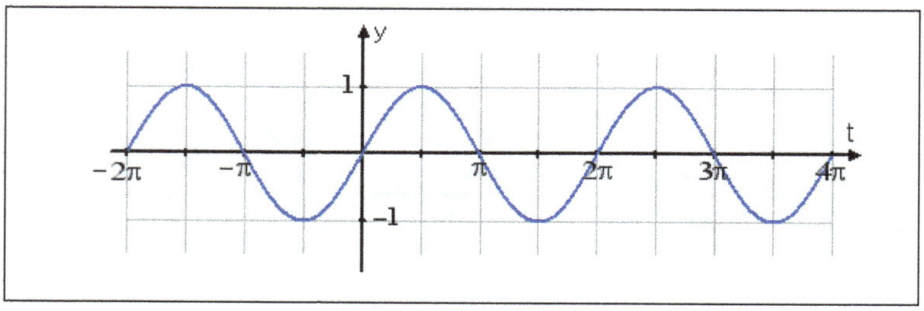

Now let's look at $k(t) = sin(t) + 3$:

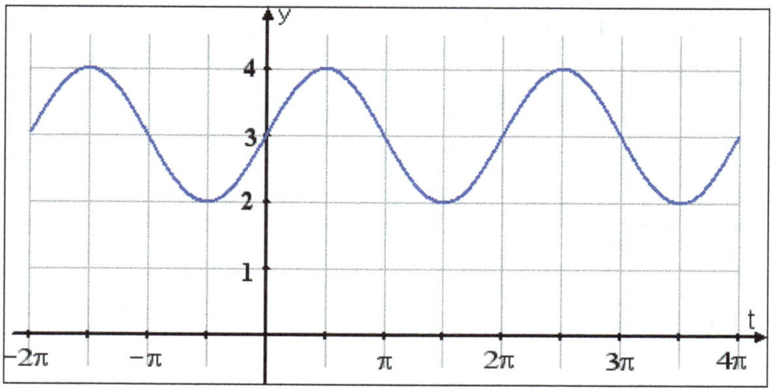

Do you see how the graph was shifted up by three units? This relationship is always true: If a number D is added outside the function, then the graph is shifted up by that number of units; if a number D is subtracted, then the graph is shifted down by that number of units.

- By what amount is the graph of $h(t) = cos(t) - 2$ shifted, and in which direction?

The trigonometric-function part is the $cos(t)$; the up-or-down shifting part is the D = −2. There's nothing else going on inside of the function, nor multiplied in front of it, so this is the regular cosine wave, but it's:

shifted downward two units

- By what amount is the graph of $t(x) = tan(x) + 0.6$ shifted, and in which direction?

The trigonometric-function part is the $tan(x)$; the up-or-down shifting part is the + 0.6. So this is the regular tangent curve, but:

shifted upward by 6/10 = 3/5 of a unit.

Putting it all together in terms of the sine wave, we have the general sine function:

$$F(t) = A\,sin(Bt - C) + D$$

where |A| is the amplitude, B gives you the period, D gives you the vertical shift (up or down), and C/B is used to find the phase shift.

Why don't we just just use C for the phase shift? Because sometimes more involved stuff is going on inside the function. Remember that the phase shift comes from what is added or subtracted *directly* to the variable. If the variable isn't alone (that is, if there's something multiplied directly on it), then there's another step to follow.

For instance, if you have something like:

$$y = \sin(2t - \pi)$$

the phase shift is *not* π units. Instead, you first have to isolate what's happening to the variable by factoring, as so:

$$y = \sin\left(2\left(-\frac{\pi}{2}\right)\right)$$

Now you can see that the phase shift will be π/2 units, not π units. So the phase shift, as a formula, is found by dividing C by B.

For $F(t) = \mathrm{A}f(\mathrm{B}t - \mathrm{C}) + \mathrm{D}$, where $f(t)$ is one of the basic trigonometric functions, we have:

- the amplitude is |A|
- the period is $\dfrac{\text{regular period}}{|\mathrm{B}|}$
- the phase shift is $\dfrac{\mathrm{C}}{\mathrm{B}}$
- the vertical shift is D

Find the amplitude, period, phase shift, and vertical shift of $s(t) = -2.5tan(4t - 3\pi) - 4$

The amplitude is given by the multipler on the trigonometric function. In this case, there's a −2.5 multiplied directly onto the tangent. This is the "A" from the formula, and tells me that the amplitude is 2.5.

The regular period for tangents is π. In this particular function, there's a 4 multiplied on the variable, so B = 4. Plugging into the period formula, we get $\frac{\pi}{4}$.

To find the phase shift, I need to isolate the variable with the shift value, so I need to factor out the 4 (also known as "C") that's multiplied on the variable. The factorization is:

$$4t - 3\pi = 4\left(t - \frac{3\pi}{4}\right)$$

Then the phase shift is $\frac{3}{4}\pi$. Because the shift value is *subtracted* from the variable, the shift is *to the right.*

The vertical shift comes from the value entirely outside of the trigonometric function; namely, the outer 4 (also known as "D", from the formula). Because this 4 is *subtracted* from the tangent, the shift will be four units *downward*from the usual center line, the x-axis.

Amplitude: 2.5

Period: $\frac{\pi}{4}$

Phase shift: to the right by $\frac{3}{4}\pi$

Vertical shift: downward by 4.

Graphs of Inverse Trigonometric Functions

Graph of Arcsin(x)

Graph of arcsine of x:

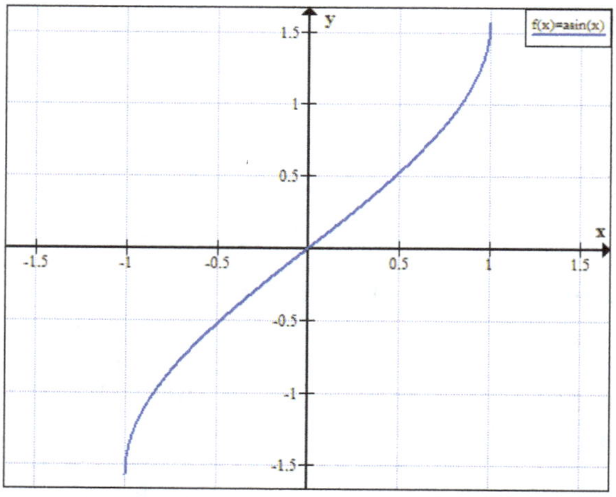

Graph, Domain and Range of Arcsin(x)

In what follows, arcsin(x) is the inverse function of f(x) = sin(x) for - π/2 ≤ x ≤ π/2. The domain of y = arcsin(x) is the range of f(x) = sin(x) for -π/2 ≤ x ≤ π/2 and is given by the interval [-1 , 1]. The range of arcsin(x) is the domain of f which is given by the interval [-π/2 , π/2]. The graph, domain and range of both f(x) = sin(x) for -π/2 ≤ x ≤ π/2 and arcsin(x) are shown below.

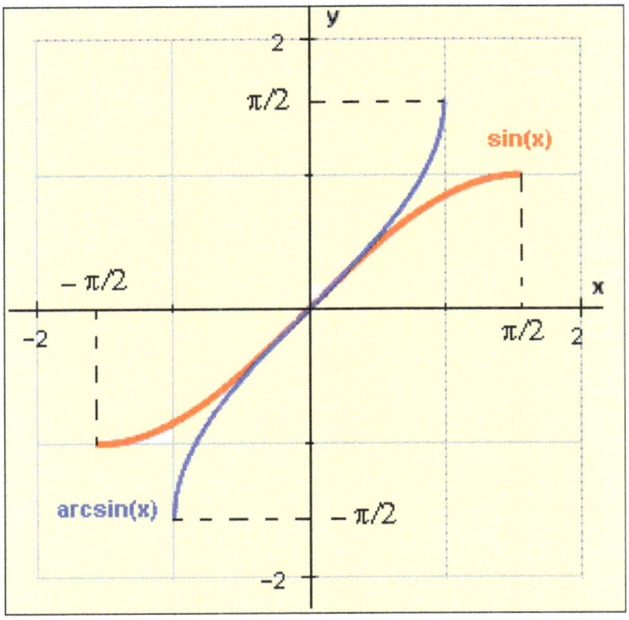

A table of values of arcsin(x) can made as follows:

x	-1	0	1
$y = \arcsin(X)$	$-\pi/2$	0	$\pi/2$

Note that there are 3 key points that may be used to graph arcsin(x). These points are: (-1,-π/2) , (0,0) and (1,π/2).

Example:

Find the domain and range of y = arcsin(x - 2) and graph it.

Solution to Example: The graph of y = arcsin(x - 2) will be that of arcsin(x) shifted 2 units to the right. The domain is found by stating that -1 ≤ x - 2 ≤ 1. Solve the double inequality to find the domain: 1 ≤ x ≤ 3

The 3 key points of arcsin(x) can also be used in this situation as follows:

$X-2$	-1	0	1
$y = \arcsin(x-2)$	$-\pi/2$	0	$\pi/2$
x	1	2	3

The value of x is calculated from the value of x - 2. For example when x - 2 = -1, solve for x to find x = 1 and so on. The domain is given by the interval [1,3] and the range is given by the interval [-π/2,π/2] The three points will now be used to graph y = arcsin(x - 2).

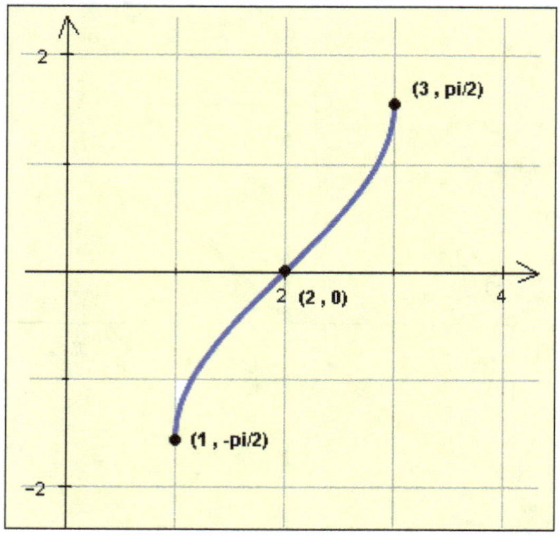

Example:

Find the domain and range of y = 2 arcsin(x + 1) and graph it.

Solution to Example: We use the 3 key points in the table as follows, then find the value 2 arcsin(x + 1) and x.

$x+1$	-1	0	1
$\arcsin(x+1)$	$-\pi/2$	0	1
$y=2\arcsin(x+1)$	$-\pi$	0	π
x	-2	-1	0

domain = [-2,0] , range = [- π , π]

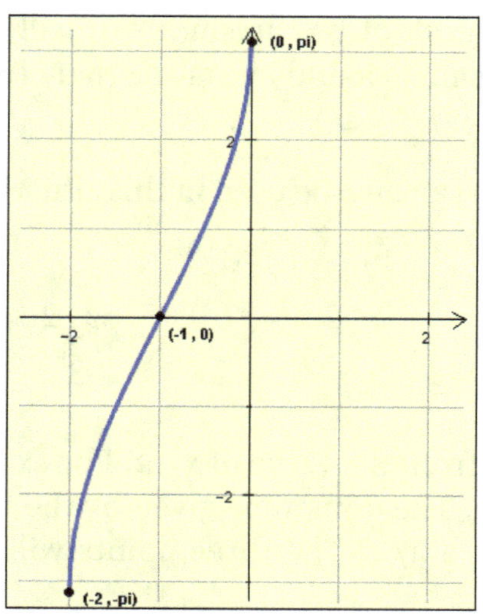

The graph is that of arcsin(x) shifted one unit to the left and stretched vertically by a factor of 2.

Example:

Find the domain and range of y = - arcsin(x - 1) and graph it.

Solution to Example: We use the 3 key points in the table as follows, then find the value - arcsin(x - 1) and x.

$x-1$	-1	0	1
$\arcsin(x-1)$	$-\pi/2$	0	$\pi/2$
$y=-\arcsin(x-1)$	$\pi/2$	0	$-\pi/2$
x	0	1	2

domain = [0 , 2] , range = [- π/2 , π/2]

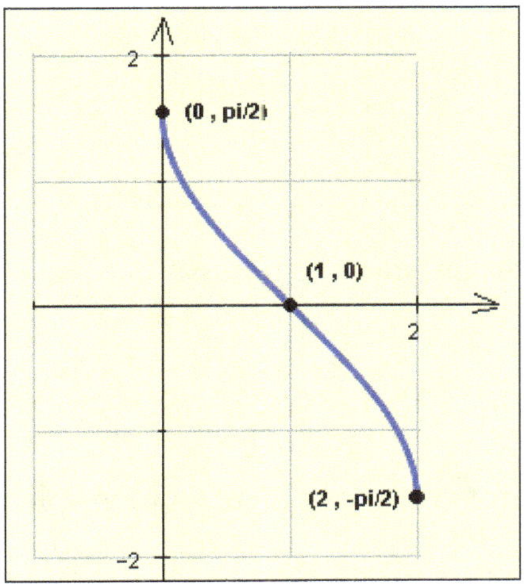

The graph is that of arcsin(x) shifted one unit to the right and reflected on the x axis.

Graph of Arccos(x)

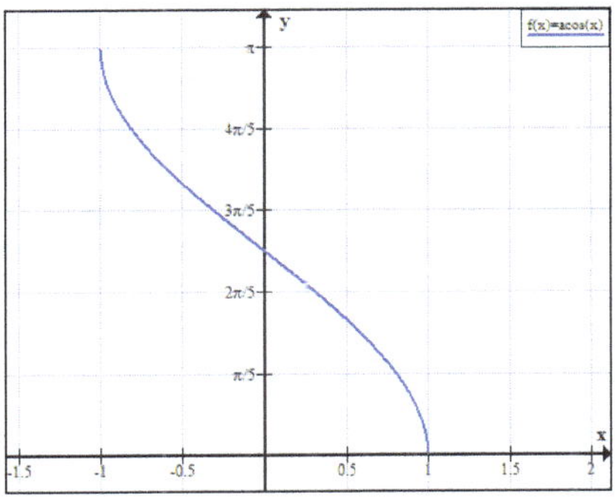

Graph, Domain and Range of Arccos(x)

In what follows, arccos(x) is the inverse function of f(x) = cos(x) for $0 \le x \le \pi$. The domain of y = arccos(x) is the range of f(x) = cos(x) for $0 \le x \le \pi$ and is given by the interval [-1 , 1]. The range of arccos(x) is the domain of f which is given by the interval [0 , π]. The graph, domain and range of both f(x) = cos(x) for $0 \le x \le \pi$ and arccos(x) are shown below.

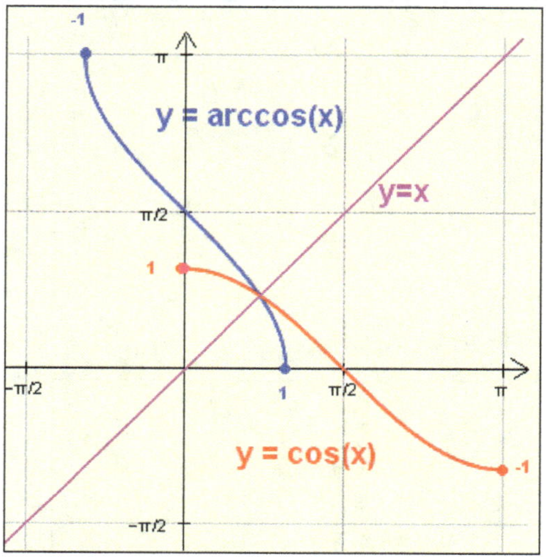

A table of values of arccos(x) can made as follows:

x	-1	0	1
$y = \arccos(x)$	π	$\pi/2$	0

Note that there are 3 key points that may be used to graph arccos(x). These points are: (-1,π) , (0,π/2) and (1,0).

Example:

Find the domain and range of y = arccos(x - 1) and graph it.

Solution to Example: The graph of y = arccos(x - 1) will be that of arccos(x) shifted 1 unit to the right. The domain is found by stating that $-1 \le x - 1 \le 1$. Solve the double inequality to find the domain: $0 \le x \le 2$

The 3 key points of arccos(x) can also be used in this situation as follows:

$x-1$	-1	0	1
$y = \arccos(x-1)$	π	$\pi/2$	0
x	0	1	2

The value of x is calculated from the value of x - 1. For example when x - 1 = -1, solve for x to find x = 0 and so on. The domain is given by the interval [0,2] and the range is given by the interval [0,π] The three points will now be used to graph y = arccos(x - 1).

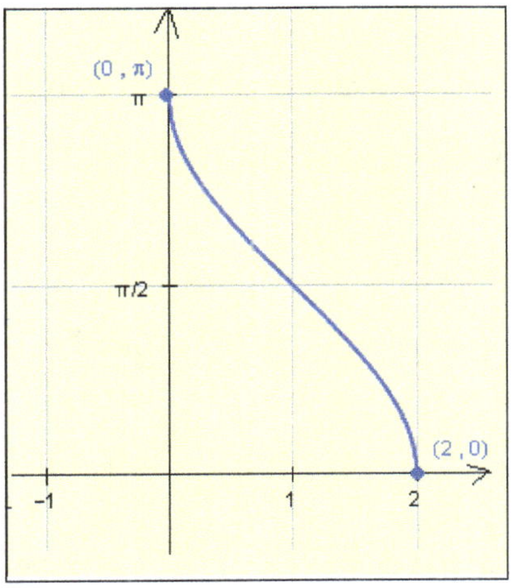

Example:

Find the domain and range of y = 2 arccos(x + 1) and graph it.

Solution to Example: We use the 3 key points in the table as follows, then find the value 2 arccos(x + 1) and x.

$x+1$	-1	0	1
$\arccos(x+1)$	π	$\pi/2$	0
$y=2\arccos(x+1)$	2π	π	0
x	-2	-1	0

domain = [-2,0] , range = [0 , 2 π]

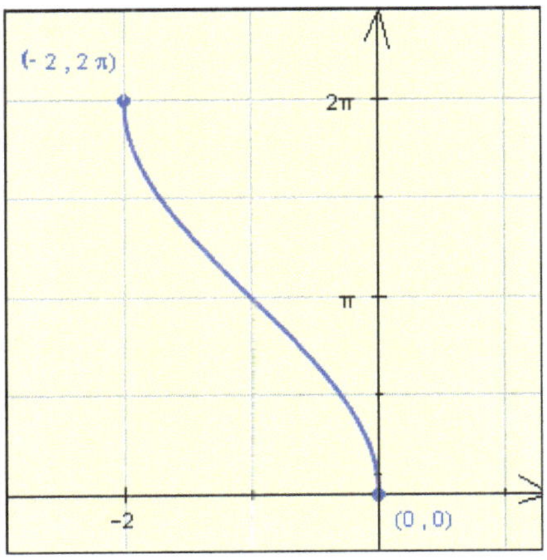

The graph is that of arccos(x) shifted one unit to the left and stretched vertically by a factor of 2.

Example:

Find the domain and range of y = - arccos(x - 1) and graph it.

Solution to Example: We use the 3 key points in the table as follows, then find the value - arccos(x - 1) and x.

$x - 1$	-1	0	1
$\arccos(x-1)$	π	$\pi/2$	0
$y = 1\arccos(x-1)$	$-\pi$	$-\pi/2$	0
x	0	1	2

domain = [0 , 2] , range = [- π , 0]

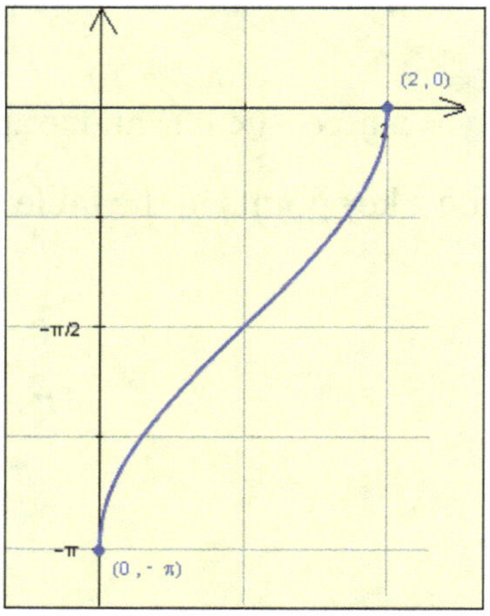

The graph is that of arccos(x) shifted one unit to the right and reflected on the x axis.

Graph of Arctan(x)

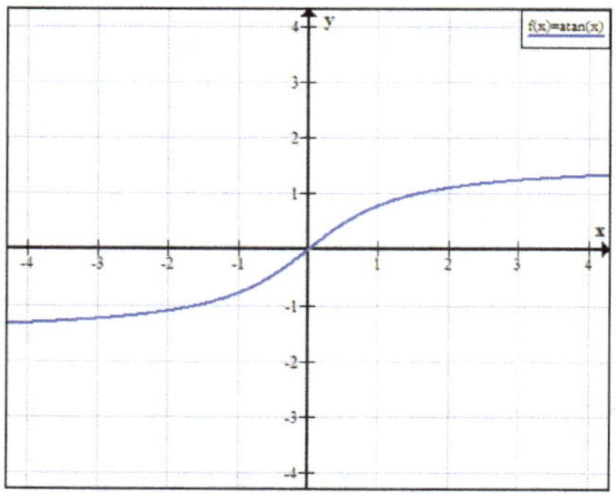

Graph of Arccsc(x)

Arc cosecant is inverse of the cosecant function.

Graph

Arc cosecant is antisymmetric function defined everywhere on real axis, ecxept the range (–1, 1) — so, function domain is $(-\infty,-1]\cup[1,+\infty)$. Its graph is depicted below.

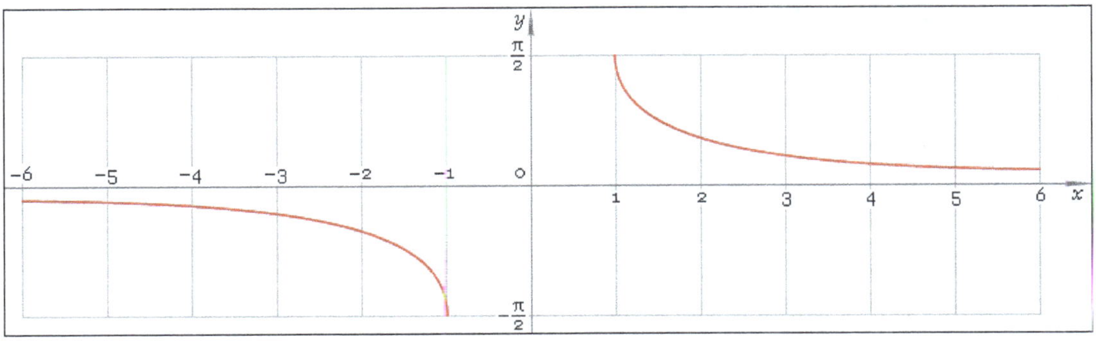

Graph of the arc cosecant function $y = arccscx$.

Function codomain is limited to the range $[-\pi/2,0)\cup(0,\pi/2]$.

Graph of Arcsec(x)

Arc secant is discontinuous function defined on entire real axis except the (–1, 1) range — so, its domain is $(-\infty,-1]\cup[1,+\infty)$. Function graph is depicted below.

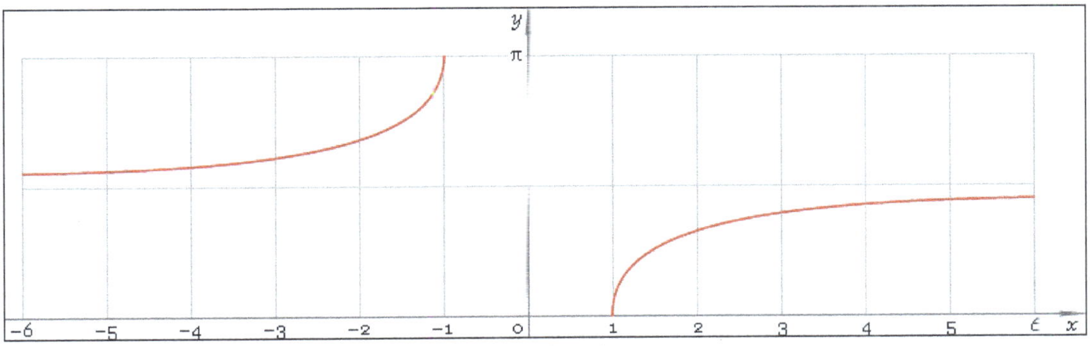

Graph of the arc secant function $y = arcsecx$.

Function codomain is limited to the range $[0,\pi/2)\cup(\pi/2,\pi]$.

Graph of Arccot (x)

Arc cotangent is monotone function defined e verywhere o n r eal a xis. I ts g raph is depicted below.

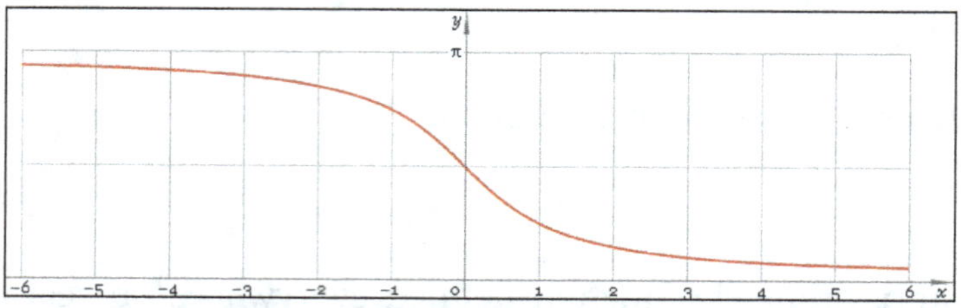

Graph of the arc cotangent function $y = arccotx$.

Function codomain is limited to the range $(0, \pi)$.

References

- Graph-of-y-equals-sin-x: math-only-math.com, Retrieved 17 July, 2019
- How-to-graph-a-tangent-function, education-math-calculus: dummies.com, Retrieved 25 August, 2019
- Graph-of-y-equals-csc-x: math-only-math.com, Retrieved 28 April, 2019
- Graph, Shape-of-Cosecant-Function: proofwiki.org, Retrieved 19 February, 2019
- Grphtrig: purplemath.com, Retrieved 18 May, 2019
- Arcsin-graph, math-trigonometry-arcsin: rapidtables.com, Retrieved 04 January, 2019
- Graphing-arcsine, Graphing: analyzemath.com, Retrieved 28 June, 2019

5

Trigonometric Identities and Laws

Pythagorean identities, double angle formulas, half-angle formulas, triple-angle formulas, etc. are studied within trigonometric identities. Law of sines, cosines, tangents, Morrie's law, De Moivre's theorem, etc. fall under the domain of trigonometric laws. The topics elaborated in this chapter will help in gaining a better perspective about these trigonometric identities and laws.

Pythagorean Identities

Pythagorean identities are identities in trigonometry that are extensions of the Pythagorean theorem. The fundamental identity states that for any angle θ,

$$\cos^2\theta + \sin^2\theta = 1.$$

Pythagorean identities are useful in simplifying trigonometric expressions, especially in writing expressions as a function of either sin or cos, as in statements of the double angle formulas.

Derivation of Fundamental Pythagorean Identity

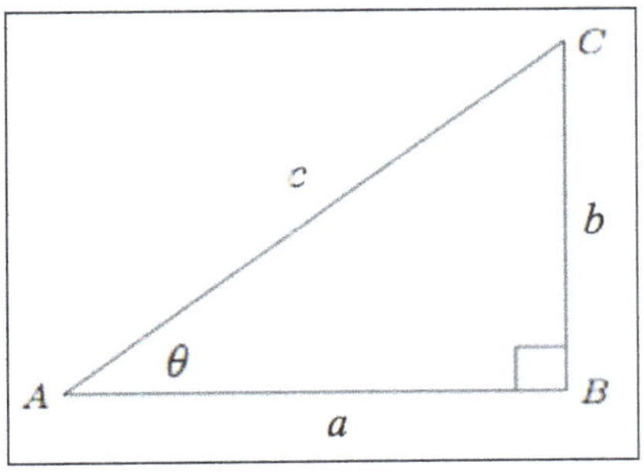

By using the Pythagorean theorem, which states that $a^2 + b^2 = c^2$, and the definitions of the basic trigonometric functions,

$$a^2 + b^2 = c^2 \Rightarrow \frac{a^2 + b^2}{c^2} = \left(\frac{a}{c}\right)^2 + \left(\frac{b}{c}\right)^2 = 1 \Rightarrow \cos^2\theta + \sin^2\theta = 1.$$

This gives the following important identity of the basic trigonometric functions.

Theorem:

Pythagorean Identity

For any angle θ, we have $\cos^2 \theta \sin^2 \theta = 1.$

Example:

If $\sin(30°) = \frac{1}{2}$, what is cos (30°)?

By the Pythagorean identity, $\sin^2(30°) + \cos^2(30°) = 1$, so

$$\cos^2(30°) = 1 - \frac{1}{4} = \frac{3}{4} \Rightarrow \cos^2(30°) = \pm\frac{\sqrt{3}}{2}.$$

However, since 30° is an acute angle, we must have $\cos(30°) = \frac{\sqrt{3}}{2}$.

Other Forms of Pythagorean Identity

From the Pythagorean Identity, we have the following corollaries:

Corollaries:

By dividing both sides by $\cos^2 \theta$, we have the following identity:

$$\frac{\cos^2\theta + \sin^2\theta}{\cos^2\theta} = \frac{1}{\cos^2\theta} \text{ or } 1 + \tan^2\theta = \sec^2\theta.$$

Similarly, by dividing both sides by $\sin^2 \theta$, we have the following identity:

$$\frac{\cos^2\theta + \sin^2\theta}{\sin^2\theta} = \frac{1}{\sin^2\theta} \text{ or } \cot^2\theta + 1 = \csc^2\theta.$$

Applications and Problem-solving

Example:

Evaluate sin 225° exactly.

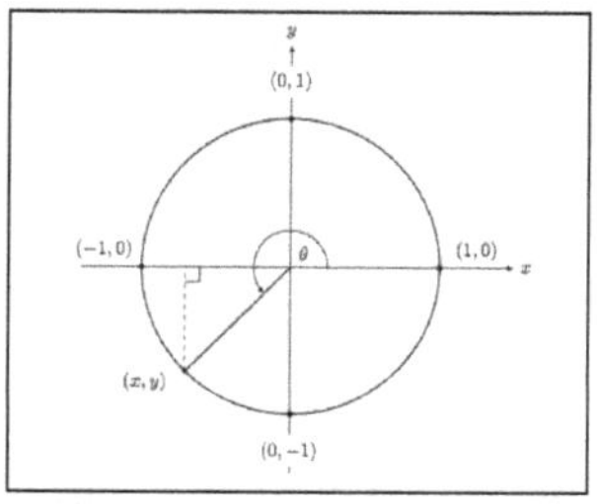

Unit circle with angle greater than 2π.

Notice that the problem reduces to finding the value of y in the above picture.

Since 225° – 180° = 45°, θ makes an angle of 45° with the negative x-axis. Further, since this is a right triangle, the angles must be 90° + 45° + 45° = 180°, which means it is isosceles. Therefore, $|x| = |y|$.

By the Pythagorean theorem,

$$\begin{aligned} x^2 + y^2 &= 1 \\ 2y^2 &= 1 \\ y^2 &= \frac{1}{2} \\ y &= \pm\frac{\sqrt{2}}{2}. \end{aligned}$$

By inspection, y is negative, so sin 225° $= \frac{\sqrt{2}}{2}$.

Most values for θ will be difficult or impossible to evaluate exactly in this way, so we often use a calculator to evaluate the approximate value of the function.

In most problems, it is best to convert all trigonometric functions into sines and cosines, in order to take advantage of the Pythagorean identities.

Example:

If $\tan\theta = \frac{1}{3}$, what is $\frac{1}{\sin^2\theta} + \frac{1}{\cos^2\theta}$?

We have

$$\tan^2\theta + 1 = \sec^2\theta = \frac{1}{\cos^2\theta}$$

$$\cot^2\theta + 1 = \csc^2\theta = \frac{1}{\sin^2\theta}.$$

Since $\tan\theta = \frac{1}{3}$, implying $\cot\theta = 3$, from $\tan^2\theta + 1 = \sec^2\theta = \frac{1}{\cos^2\theta}$ and $\cot^2\theta + 1 = \csc^2\theta = \frac{1}{\sin^2\theta}$ we have

$$\frac{1}{\sin^2\theta} + \frac{1}{\cos^2\theta} = \cot^2\theta + 1 + \tan^2\theta + 1$$

$$= 3^2 + 1 + \left(\frac{1}{3}\right)^2 + 1$$

$$= \frac{100}{9}.$$

The Pythagorean identities are so useful that it is often worth squaring expressions to take advantage of them:

Example:

If $\sin\theta + \cos\theta = \frac{1}{2}$, what is $\sin\theta \cdot \cos\theta$?

$$(\sin\theta + \cos\theta)^2 = \left(\frac{1}{2}\right)^2$$

$$\sin^2\theta + 2\cdot\sin\theta\cdot\cos\theta + \cos^2\theta = \frac{1}{4}.$$

Since $\sin^2\theta + \cos^2\theta = 1$, it follows that:

$$\sin^2\theta + 2\cdot\sin\theta\cdot\cos\theta + \cos^2\theta = \frac{1}{4}$$

$$1 + 2\cdot\sin\theta\cdot\cos\theta = \frac{1}{4}$$

$$2\cdot\sin\theta\cdot\cos\theta = \frac{1}{4} - 1$$

$$\Rightarrow \sin\theta\cdot\cos\theta = -\frac{3}{8}.$$

Example:

If $\tan\theta + \frac{1}{\tan\theta} = 8$, where $0 < \theta < \frac{\pi}{2}$, what is $\sin\theta + \cos\theta$?

We have

$$\tan\theta + \frac{1}{\tan\theta} = 8$$

$$\frac{\sin\theta}{\cos\theta} + \frac{\cos\theta}{\sin\theta} = 8$$

$$\frac{\sin^2\theta + \cos^2\theta}{\sin\theta\cdot\cos\theta} = 8$$

$$\frac{\sin^2\theta+\cos^2\theta}{\sin\theta\cdot\cos\theta}=8$$

$$\frac{1}{\sin\theta\cdot\cos\theta}=8$$

$$\sin\theta\cdot\cos\theta=\frac{1}{8}.$$

Since $\sin\theta\cdot\cos\theta=\frac{1}{8}$, thus we have

$$(\sin\theta+\cos\theta)^2=1+2\cdot\sin\theta\cdot\cos\theta$$

$$=1+2\times\frac{1}{8}$$

$$=\frac{5}{4}.$$

Since θ lies in the interval $\left(0,\frac{\pi}{2}\right)$, the value of $\sin\theta+\cos\theta$ is positive. Thus,

$$\sin\theta+\cos\theta=\sqrt{\frac{5}{4}}=\frac{\sqrt{5}}{2}.$$

Example:

If the two roots of the equation $2x^2+px-1=0$ are sin θ and cos θ, what is p?

From Vieta's formula, we have

$$\sin\theta+\cos\theta=-\frac{p}{2}$$

$$\sin\theta\cdot\cos\theta=-\frac{1}{2}.$$

Squaring both sides of $\sin\theta+\cos\theta=-\frac{p}{2}$, we have:

$$\sin^2\theta+2\cdot\sin\theta\cdot\cos\theta+\cos^2\theta=\frac{p^2}{4}$$

$$1+2\cdot\sin\theta\cdot\cos\theta=\frac{p^2}{4}.$$

Substituting $\sin\theta\cdot\cos\theta=-\frac{1}{2}$ to $1+2\cdot\sin\theta\cdot\cos\theta=\frac{p^2}{4}$, we have,

$$1+2\cdot\left(-\frac{1}{2}\right)=\frac{p^2}{4}$$
$$\frac{p^2}{4}=0$$
$$\Rightarrow p=0.$$

Trigonometric Addition Formulas

Angle addition formulas express trigonometric functions of sums of angles $\alpha\pm\beta$ in terms of functions of α and β. The fundamental formulas of angle addition in trigonometry are given by:

$$\sin(\alpha+\beta)=\sin\alpha\cos\beta+\sin\beta\cos\alpha$$
$$\sin(\alpha-\beta)=\sin\alpha\cos\beta-\sin\beta\cos\alpha$$
$$\cos(\alpha+\beta)=\cos\alpha\cos\beta-\sin\alpha\sin\beta$$
$$\cos(\alpha-\beta)=\cos\alpha\cos\beta+\sin\alpha\sin\beta$$
$$\tan(\alpha+\beta)=\frac{\tan\alpha+\tan\beta}{1-\tan\alpha\tan\beta}$$
$$\tan(\alpha-\beta)=\frac{\tan\alpha-\tan\beta}{1+\tan\alpha\tan\beta}.$$

The first four of these are known as the prosthaphaeresis formulas, or sometimes as Simpson's formulas.

The sine and cosine angle addition identities can be compactly summarized by the matrix equation:

$$\begin{bmatrix}\cos\alpha & \sin\alpha\\ -\sin\alpha & \cos\alpha\end{bmatrix}\begin{bmatrix}\cos\beta & \sin\beta\\ -\sin\beta & \cos\beta\end{bmatrix}=\begin{bmatrix}\cos(\alpha+\beta) & \sin(\alpha+\beta)\\ -\sin(\alpha+\beta) & \cos(\alpha+\beta)\end{bmatrix}.$$

These formulas can be simply derived using complex exponentials and the Euler formula as follows.

$$\begin{aligned}\cos(\alpha+\beta)+i\sin(\alpha+\beta)&=e^{i(\alpha+\beta)}\\ &=e^{i\alpha}e^{i\beta}\\ &=(\cos\alpha+i\sin\alpha)(\cos\beta+i\sin\beta)\\ &=(\cos\alpha\cos\beta-\sin\alpha\sin\beta)+i(\sin\alpha\cos\beta+\cos\alpha\sin\beta).\end{aligned}$$

Equating real and imaginary parts then gives $\sin(\alpha+\beta)=\sin\alpha\cos\beta+\sin\beta\cos\alpha$ and $\cos(\alpha+\beta)=\cos\alpha\cos\beta-\sin\alpha\sin\beta$, and $\sin(\alpha-\beta)=\sin\alpha\cos\beta-\sin\beta\cos\alpha$ and $\cos(\alpha-\beta)=\cos\alpha\cos\beta+\sin\alpha\sin\beta$ follow immediately by substituting $-\beta$ for β.

Taking the ratio of $\sin(\alpha+\beta)=\sin\alpha\cos\beta+\sin\beta\cos\alpha$ and $\cos(\alpha+\beta)$ $=\cos\alpha\cos\beta-\sin\alpha\sin\beta$ gives the tangent angle addition formula:

$$\begin{aligned}
\tan(\alpha+\beta) &\equiv \frac{\sin(\alpha+\beta)}{\cos(\alpha+\beta)} \\
&= \frac{\sin\alpha\cos\beta+\sin\beta\cos\alpha}{\cos\alpha\cos\beta-\sin\alpha\sin\beta} \\
&= \frac{\dfrac{\sin\alpha}{\cos\alpha}+\dfrac{\sin\beta}{\cos\beta}}{1-\dfrac{\sin\alpha\sin\beta}{\cos\alpha\cos\beta}} \\
&= \frac{\tan\alpha+\tan\beta}{1-\tan\alpha\tan\beta}
\end{aligned}$$

The double-angle formulas are:

$$\begin{aligned}
\sin(2\alpha) &= 2\sin\alpha\cos\alpha \\
\cos(2\alpha) &= \cos^2\alpha-\sin^2\alpha \\
&= 2\cos^2\alpha-1 \\
&= 1-2\sin^2\alpha \\
\tan(2\alpha) &= \frac{2\tan\alpha}{1-\tan^2\alpha}.
\end{aligned}$$

Multiple-angle formulas are given by:

$$\sin(n,x)=\sum_{k=0}^{n}\binom{n}{k}\cos^k x\sin^{n-k} x\sin\left[\frac{1}{2}(n-k)\pi\right]$$

$$\cos(n,x)=\sum_{k=0}^{n}\binom{n}{k}\cos^k x\sin^{n-k} x\cos\left[\frac{1}{2}(n-k)\pi\right]$$

and can also be written using the recurrence relations:

$$\sin(nx) = 2\sin\left[(n-1)x\right]\cos x - \sin\left[(n-k)x\right]$$
$$\cos(nx) = 2\cos\left[(n-1)x\right]\cos x - \cos\left[(n-k)x\right],$$

$$\tan(nx) = \frac{\tan\left[(n-1)x\right] + \tan x}{1 - \tan\left[(n-1)x\right]\tan x}.$$

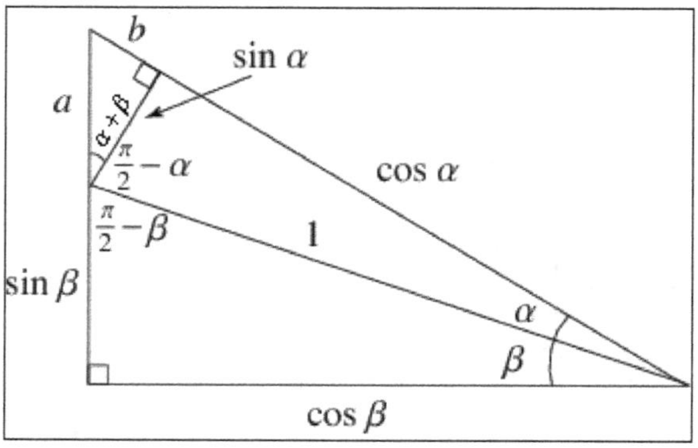

The angle addition formulas can also be derived purely algebraically without the use of complex numbers. Consider the small right triangle in the figure above, which gives:

$$a = \frac{\sin\alpha}{\cos(\alpha+\beta)}$$
$$b = \sin\alpha\tan(\alpha+\beta)$$

Now, the usual trigonometric definitions applied to the large right triangle give:

$$\sin(\alpha+\beta) = \frac{\sin\beta + a}{\cos a + b}$$
$$= \frac{\sin\beta + \dfrac{\sin a}{\cos(\alpha+\beta)}}{\cos\alpha + \sin\alpha\dfrac{\sin(\alpha+\beta)}{\cos(\alpha+\beta)}}$$

$$\cos(\alpha+\beta) = \frac{\cos\beta}{\cos a + b}$$
$$= \frac{\cos\beta}{\cos\alpha + \sin\alpha\dfrac{\sin(\alpha+\beta)}{\cos(\alpha+\beta)}}.$$

Solving these two equations simultaneously for the variables $\sin(\alpha+\beta)$ and $\cos(\alpha+\beta)$ then immediately gives:

$$\sin(\alpha+\beta)=\frac{\cos\alpha\sin\alpha+\cos\beta\sin\beta}{\cos\alpha\cos\beta+\sin\alpha\sin\beta}$$

$$\cos(\alpha+\beta)=\frac{\cos^2\beta-\sin^2\alpha}{\cos\alpha\cos\beta+\sin\alpha\sin\beta}.$$

These can be put into the familiar forms with the aid of the trigonometric identities:

$$(\cos\alpha\cos\beta+\sin\alpha\sin\beta)(\sin\alpha\cos\beta+\sin\beta\cos\alpha)=\cos\beta\sin\beta+\cos\alpha\sin\alpha$$

and:

$$\begin{aligned}(\cos\alpha\cos\beta+\sin\alpha\sin\beta)(\cos\alpha\cos\beta-\sin\alpha\sin\beta)&=\cos^2\alpha\cos^2\beta-\sin^2\alpha\sin^2\beta\\&=1-\sin^2\alpha-\sin^2\beta\\&=\cos^2\alpha-\sin^2\beta\\&=\cos^2\beta-\sin^2\alpha,\end{aligned}$$

which can be verified by direct multiplication:

$$\sin(\alpha+\beta)=\sin\alpha\cos\beta+\sin\beta\cos\alpha$$

$$\cos(\alpha+\beta)=\cos\alpha\cos\beta-\sin\alpha\cos\beta,$$

as before.

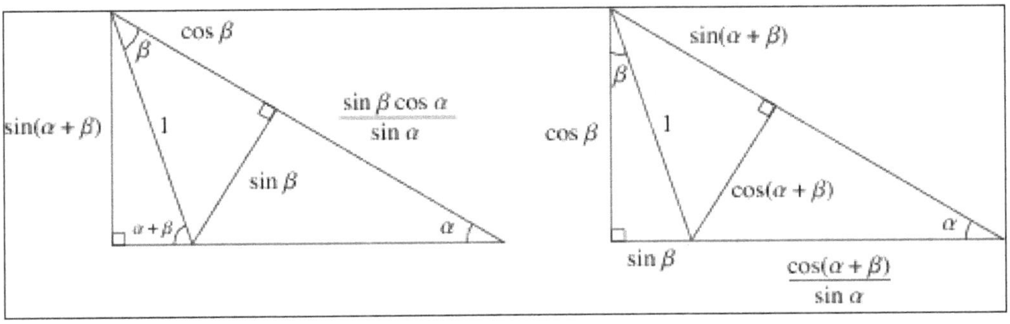

A similar proof due to Smiley and Smiley uses the left figure above to obtain:

$$\sin\alpha=\frac{\sin(\alpha+\beta)}{\cos\beta+\dfrac{\sin\beta\cos\alpha}{\sin\alpha}},$$

from which it follows that:

$$\sin(\alpha+\beta)=\sin\alpha\cos\beta+\sin\beta\cos\alpha.$$

Similarly, from the right figure,

$$\frac{\sin\alpha}{\cos\alpha}=\frac{\cos\beta}{\sin\beta+\frac{\cos(\alpha+\beta)}{\sin\alpha}},$$

So,

$$\cos(\alpha+\beta)=\cos\alpha\cos\beta-\sin\alpha\sin\beta.$$

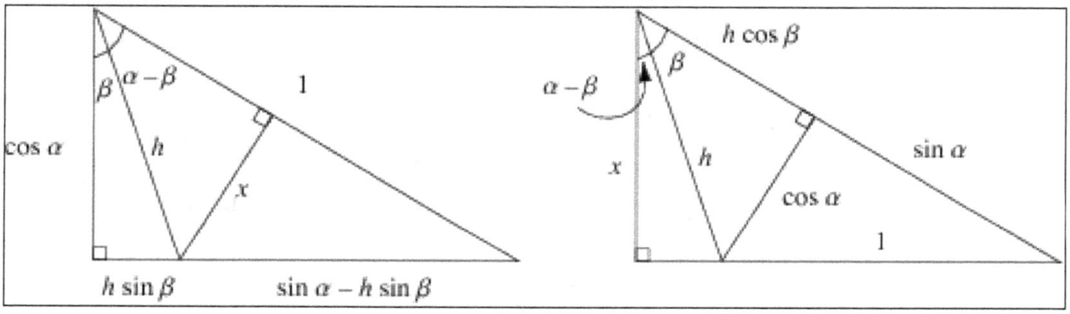

Similar diagrams can be used to prove the angle subtraction formulas. In the figure at left,

$$\begin{aligned}h&=\frac{\cos\alpha}{\cos\beta}\\x&=h\sin(\alpha-\beta)\\&=(\sin\alpha-h\sin\beta)\cos\alpha\end{aligned}$$

Giving,

$$\sin(\alpha-\beta)=\sin\alpha\cos\beta-\cos\alpha\sin\beta.$$

Similarly, in the figure at right,

$$\begin{aligned}h&=\frac{\cos\alpha}{\sin\beta}\\x&=h\cos(\alpha-\beta)\\&=(\sin\alpha+h\cos\beta)\cos\alpha\end{aligned}$$

Giving,

$$\cos(\alpha-\beta)=\cos\alpha\cos\beta+\sin\alpha\sin\beta$$

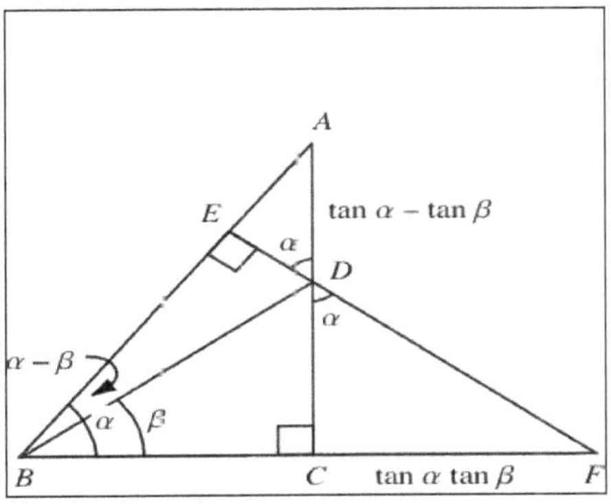

A more complex diagram can be used to obtain a proof from the $\tan(\alpha-\beta)$ identity. In the above figure, let $BF/BE = AD/DE$. Then:

$$\tan(\alpha-\beta)=\frac{DE}{BE}=\frac{AD}{BF}=\frac{\tan\alpha-\tan\beta}{1+\tan\alpha\tan\beta}.$$

An interesting identity relating the sum and difference tangent formulas is given by:

$$\begin{aligned}\frac{\tan(\alpha-\beta)}{\tan(\alpha+\beta)}&=\frac{\sin(\alpha-\beta)\cos(\alpha+\beta)}{\cos(\alpha-\beta)\sin(\alpha+\beta)}\\&=\frac{(\cos\alpha\cos\beta-\sin\beta\cos\alpha)(\cos\alpha\cos\beta-\sin\alpha\sin\beta)}{(\cos\alpha\cos\beta+\sin\alpha\sin\beta)(\sin\alpha\cos\beta+\sin\beta\cos\alpha)}\\&=\frac{\sin\alpha\cos\alpha-\sin\beta\cos\beta}{\sin\alpha\cos\alpha+\sin\beta\cos\beta}.\end{aligned}$$

Double-angle Formulae

Formulas expressing trigonometric functions of an angle $2x$ in terms of functions of an angle x,

$$\begin{aligned}\sin(2x)&=2\sin x\cos x\\\cos(2x)&=\cos^2 x-\sin^2 x\\&=2\cos^2 x-1\\&=1-2\sin^2 x\end{aligned}$$

$$\tan(2x) = \frac{2\tan x}{1-\tan^2 x}.$$

The corresponding hyperbolic function double-angle formulas are:

$$\sinh(2x) = 2\sinh x \cosh x$$
$$\cosh(2x) = 2\cosh^2 x - 1$$
$$\tanh(2x) = \frac{2\tanh x}{1+\tanh^2 x}$$

Double-angle Formula for Sine

$$\sin 2\theta = 2\sin\theta\cos\theta$$

where sin and cos denote sine and cosine respectively.

Corollary:

$$\sin 2\theta = \frac{2\tan\theta}{1+\tan^2\theta}$$

Proof:

$$\cos 2\theta + i\sin 2\theta = (\cos\theta + i\sin\theta)^2 \qquad \text{De Moivre's Formula}$$
$$= \cos^2\theta + i^2\sin^2\theta + 2i\cos\theta\sin\theta$$
$$= \cos^2\theta - \sin^2\theta + 2i\cos\theta\sin\theta$$
$$\sin 2\theta = 2\cos\theta\sin\theta \qquad \text{equating imaginary parts}$$

Proof:

$$\sin 2\theta = \sin(\theta+\theta)$$
$$= \sin\theta\cos\theta + \cos\theta\sin\theta \qquad \text{Sine of Sum}$$
$$2\sin\theta\cos\theta$$

Proof:

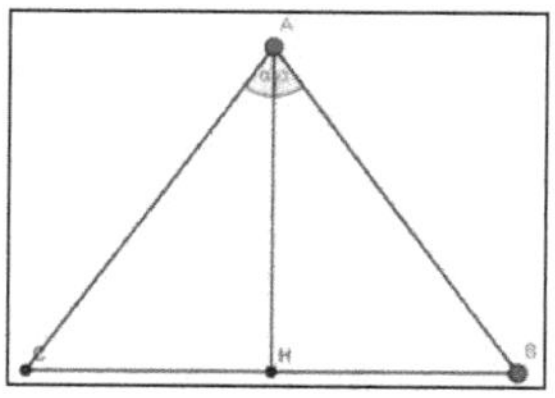

Consider a Isosceles Triangle ΔABC with base BC, and head angle $\angle$BAC = 2α.

Draw a angle bisector to $\angle$BAC and name it AH.

$$\angle BAH = \angle CAH = \alpha$$

From Angler Bisector and Altitude coincide iff triangle is isosceles:

$$AH \perp BC.$$

From Area of Triangle in Terms of Two Sides and Angle:

$$\text{Area}\left(\Delta BAH\right) = \frac{BA \cdot AH \sin\alpha}{2}$$

$$\text{Area}\left(\Delta CAH\right) = \frac{CA \cdot AH \sin\alpha}{2}$$

By definition of sine:

$$AH = CA\cos\alpha$$

$$AH = BA\cos\alpha$$

And so:

$$\text{Area}\left(\Delta BAH\right) = \frac{BA \cdot CA\cos\alpha\sin\alpha}{2}$$

$$\text{Area}\left(\Delta CAH\right) = \frac{CA \cdot BA\cos\alpha\sin\alpha}{2}$$

$$\begin{aligned}\text{Area}\left(\Delta ABC\right) &= \text{Area}\left(\Delta BAH\right) + Area\left(\Delta CAH\right)\\ &= \frac{BA \cdot CA\cos\alpha\sin\alpha}{2} + \frac{CA \cdot BA\cos\alpha\sin\alpha}{2}\\ &= BAC\,A\cos\alpha\sin\alpha\end{aligned}$$

$= \dfrac{BA \cdot CA\sin 2\alpha}{2}$ Area of Triangle in Terms of Two Sides and Angle (ΔABC)

And by cancelling out common terms:

$$\sin 2\alpha = 2\cos\alpha\sin\alpha$$

Proof:

$$\sin 2\theta = \frac{1}{2i}\left(e^{2i\theta} - e^{-2i\theta}\right)$$ Sine Exponential Formulation

$= \frac{1}{2i}\left(e^{i\theta} + e^{i\theta}\right)\left(e^{i\theta} - e^{-i\theta}\right)$ Difference of Two Squares

$$= 2\left(\frac{e^{i\theta} - e^{-i\theta}}{2i} \cdot \frac{e^{i\theta} + e^{-i\theta}}{2}\right)$$

$= 2\sin\theta\cos\theta$ Sine Exponential Formulation, Cosine Exponential Formulation

Double-angle Formula for Cosine

Theorem:

$$\cos 2\theta = \cos^2\theta - \sin^2\theta$$

where cos and sin denote cosine and sine respectively.

Corollary 1:

$$\cos 2\theta = 2\cos^2\theta - 1$$

Corollary 2:

$$\cos 2\theta = 1 - 2\sin^2\theta$$

Corollary 3:

$$\cos 2\theta = \frac{1 + \tan^2\theta}{1 + \tan^2\theta}$$

Proof:

$\cos 2\theta + i\sin 2\theta = \left(\cos\theta + i\sin\theta\right)^2$ De Moivre's Formula

$= \cos^2\theta + i^2\sin^2\theta + 2i\cos\theta\sin\theta$

$= \cos^2\theta - \sin^2\theta + 2i\cos\theta\sin\theta$

$\cos 2\theta = \cos^2 = \theta - \sin^2\theta$ equating real parts

Proof:

$\cos 2\theta = \cos\left(\theta + \theta\right)$

$= \cos\theta\cos\theta - \sin\theta\sin\theta$ Cosine of Sum

$= \cos^2\theta - \sin^2\theta$

Proof:

Starting from the right, we have:

$$\cos^2\theta - \sin^2\theta = \left(\frac{1}{2}\left(e^{i\theta}+e^{-i\theta}\right)\right)^2 - \left(\frac{1}{2i}\left(e^{i\theta}-e^{-i\theta}\right)\right)^2$$ Cosine Exponential Formulation, Sine Exponential Formulation

$$=\frac{1}{4}\left(e^{i\theta}+e^{-i\theta}\right)^2+\frac{1}{4}\left(e^{i\theta}-e^{-i\theta}\right)^2$$ i is the imaginary unit

$$=\frac{1}{4}\left(e^{2i\theta}+2+e^{-2i\theta}+e^{2i\theta}-2+e^{-2i\theta}\right)$$ Square of Sum, Square of Difference

$$=\frac{1}{2}\left(e^{2i\theta}+e^{-2i\theta}\right)$$ Simplifying

$$=\cos 2\theta$$ Cosine Exponential Formulation

Proof:

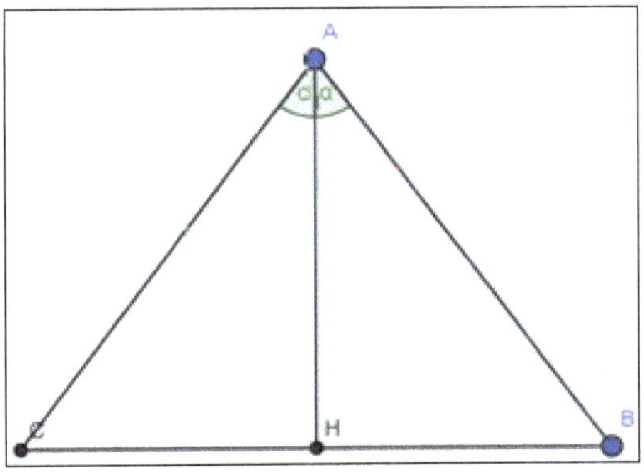

Consider a Isosceles Triangle ΔABC with base BC, and head angle ∠BAC = 2α.

Draw a angle bisector to $\angle BAC$ and name it AH.

$$\angle BAH = \angle CAH = \alpha$$

From Angler Bisector and Altitude coincide iff triangle is isosceles:

$$AH \perp BC.$$

From Law of Cosines:

$$CB^2 = AB^2 + AB^2 - 2\cdot AB \cdot AC \cdot \cos 2\alpha$$

From Pythagoras's Theorem:

$$AC^2 = CH^2 + AH^2 \qquad \text{In triangle } \Delta AHC$$

$$\Rightarrow CH^2 = AC^2 - AH^2$$

$$AB^2 = BH^2 + AH^2 \qquad \text{In triangle } \Delta AHB$$

$$\Rightarrow BH^2 = BC^2 - AH^2$$

By definition of sin:

$$CH = AC\sin\alpha$$

$$BH = AB\sin\alpha$$

By definition of cos:

$$AH = AB\cos\alpha = AC\cos\alpha$$

And so:

$$AH^2 = AB\cdot AC\cdot\cos^2\alpha$$

$$CH^2 = AC^2 - AH^2 \qquad (2.1)$$

$$= AC^2 - AB\cdot AC\cdot\cos^2\alpha \qquad \text{assigning}(4)$$

Now:

$$CB^2 = (CH + BH)^2$$

$$= CH^2 + BH^2 + 2\cdot CH\cdot BH \quad \text{Square of Sum}$$

$$= AC^2 - AB\cdot AC\cdot\cos^2\alpha + AB^2 - AB\cdot AC\cdot\cos^2\alpha + 2\cdot CH\cdot BH \qquad \text{assigning}$$
(5.1),(5.2)

$$= AC^2 + AB^2 - 2\cdot AB\cdot AC\cdot\cos^2\alpha + 2\cdot CH\cdot BH \qquad \text{simplifying}$$

$$= AC^2 + AB^2 - 2\cdot AB\cdot AC\cdot\cos^2\alpha + 2\cdot AB\cdot AC\cdot\sin^2\alpha \qquad \text{assigning}$$
(3.1),(3.2)

$$= AC^2 + AB^2 - 2\cdot AB\cdot AC\left(\cos^2\alpha - \sin^2\alpha\right) \quad \text{simplifying}$$

$$= AC^2 + AB^2 - 2\cdot AB\cdot AC\cdot\cos 2\alpha$$

And so we get the equation:

$$AC^2 + AB^2 - 2 \cdot AB \cdot AC\left(\cos^2 \alpha - \sin^2 \alpha\right) = AC^2 + AB^2 - 2 \cdot AB \cdot AC \cdot \cos 2\alpha$$
$$\Rightarrow \cos^2 \alpha - \sin^2 \alpha = \cos 2\alpha$$

Double-angle Formula for Tangent

Theorem:

$$\tan(2\theta) = \frac{2\tan\theta}{1-\tan^2\theta}$$

where tan denotes tangent.

Proof:

$$\tan(2\theta) = \frac{\sin(2\theta)}{\cos(2\theta)}$$ Tangent is Sine divided by Cosine

$$= \frac{2\cos\theta\sin\theta}{\cos^2\theta - \sin^2\theta}$$ Double Angle Formula for Sine and Double Angle Formula for Cosine

$$= \frac{\dfrac{2\cos\theta\sin\theta}{\cos^2\theta}}{\dfrac{\cos^2\theta - \sin^2\theta}{\cos^2\theta}}$$ dividing top and bottom by $\cos^2\theta$

$$= \frac{2\tan\theta}{1-\tan^2\theta}$$ Simplifying: Tangent is Sine divided by Cosine

Proof:

$$\tan(2\theta) = \tan(\theta+\theta)$$

$$= \frac{\tan\theta + \tan\theta}{1-\tan\theta\tan\theta}$$ Tangent of Sum

$$= \frac{2\tan\theta}{1-\tan^2\theta}$$

Proof:

$$\frac{2\tan\theta}{1-\tan^2\theta} = \frac{2i\dfrac{1-e^{2i\theta}}{1+e^{2i\theta}}}{1-\left(i\dfrac{1-e^{2i\theta}}{1+e^{2i\theta}}\right)^2}$$ Tangent Exponential Formulation

$$=\frac{2i\left(1-e^{2i\theta}\right)\left(1+e2i\theta\right)}{\left(1+e^{2i\theta}\right)^2+\left(1+e^{2i\theta}\right)^2}$$ multiplying both numerator and denominator by $\left(1+e^{2i\theta}\right)^2$; i is the imaginary unit

$$=\frac{2i\left(1-e^{4i\theta}\right)}{1+2e^{2i\theta}+e^{4i\theta}+1-2e^{2i\theta}+e^{4i\theta}}$$ Difference of Two Squares, Square of Sum, Square of Difference

$$=\frac{2i\left(1-e^{4i\theta}\right)}{2+2e^{4i\theta}}$$ simplifying

$$=i\frac{1-e^{4i\theta}}{2+2e^{4i\theta}}$$ simplifying

$$=\tan\left(2\theta\right)$$ Tangent Exponential Formulation

Half-angle Formulae

Half-angle formulas and formulas expressing trigonometric functions of an angle x/2 in terms of functions of an angle x. For real x,

$$\sin\left(\frac{1}{2}x\right)=\left(-1\right)^{\lfloor x/2(2\pi)\rfloor}\sqrt{\frac{1-\cos x}{2}}$$

$$\cos\left(\frac{1}{2}x\right)=\left(-1\right)^{\lfloor(x-\pi)/(2\pi)\rfloor}\sqrt{\frac{1+\cos x}{2}}$$

$$\tan\left(\frac{1}{2}x\right)=\left(-1\right)^{\lfloor x/\pi\rfloor}\sqrt{\frac{1-\cos x}{1+\cos x}}$$

$$=\frac{\sin x}{1+\cos x}$$

$$=\frac{1-\cos x}{\sin x}$$

$$=\frac{\tan x\sin x}{\tan x+\sin x}$$

$$=\frac{\left(-1\right)^{\lfloor(x+\pi/2)/\pi\rfloor}\sqrt{1+\tan^2 x}-1}{\tan x}$$

The corresponding hyperbolic function half-angle formulas are:

$$\sinh\left(\frac{1}{2}x\right) = \operatorname{sgn}(x)\sqrt{\frac{\cosh x - 1}{2}}$$

$$\cosh\left(\frac{1}{2}x\right) = \sqrt{\frac{\cosh x + 1}{2}}$$

$$\tanh\left(\frac{1}{2}x\right) = \frac{\sinh x}{\cosh x + 1} = \frac{\cosh x - 1}{\sinh x}.$$

The Weierstrass substitution makes use of the half-angle formulas:

$$\cos t = \frac{1 - \tan^2\left(\frac{1}{2}t\right)}{1 + \tan^2\left(\frac{1}{2}t\right)}$$

$$\sin t = \frac{2\tan\left(\frac{1}{2}t\right)}{1 + \tan^2\left(\frac{1}{2}t\right)}.$$

Half-angle Formula for Sine

Theorem:

$$\sin\frac{\theta}{2} = +\sqrt{\frac{1-\cos\theta}{2}} \quad \text{for } \frac{\theta}{2} \quad \text{in quadrant I or quadrant II}$$

$$\sin\frac{\theta}{1} = -\sqrt{\frac{1-\cos\theta}{2}} \quad \text{for } \frac{\theta}{2} \quad \text{in quadrant III or quadrant IV}$$

where sin denotes sine and cos denotes cosine.

Proof:

$$\cos\theta = 1 - 2\sin^2\frac{\theta}{2} \qquad \text{Double Angle Formula for Cosine: Corollary 2}$$

$$2\sin^2\frac{\theta}{1} = 1 - \cos\theta$$

$$\sin\frac{\theta}{2} = \pm\sqrt{\frac{1-\cos\theta}{2}}$$

We also have that:

In quadrant I and quadrant II, sin θ > 0

In quadrant III and quadrant IV, sin θ < 0.

Proof:

Define:

$$u = \frac{\theta}{2}$$

Then:

$$\sin^2 u = \frac{1 - \cos 2u}{2} \qquad \text{Power Reduction Formulas}$$

$$\sin\frac{\theta}{2} = \pm\sqrt{\frac{1-\cos\theta}{2}}$$

We also have that:

In quadrant I, and quadrant II, sin θ > 0

In quadrant III, and quadrant IV, sin θ < 0.

Half Angle Formula for Cosine

Theorem:

$$\cos\frac{\theta}{2} = +\sqrt{\frac{1+\cos\theta}{2}}, \quad \text{for } \frac{\theta}{2}, \text{ in quadrant I or quadrant IV}$$

$$\cos\frac{\theta}{2} = -\sqrt{\frac{1+\cos\theta}{2}}, \quad \text{for } \frac{\theta}{2}, \text{ in quadrant II or quadrant III}$$

where cos denotes cosine.

Proof:

$$\cos\theta = 2\cos^2\frac{\theta}{2} - 1 \qquad \text{Double Angle Formula for Cosine: Corollary}$$

$$2\cos^2\frac{\theta}{2} = 1 + \cos\theta$$

$$\cos\frac{\theta}{2} = \pm\sqrt{\frac{1+\cos\theta}{2}}$$

We also have that:

In quadrant I, and quadrant IV, $\cos\frac{\theta}{2} > 0$

In quadrant II and quadrant III, $\cos\frac{\theta}{2} < 0$.

Proof:

Define:

$$u = \frac{\theta}{2}$$

Then:

$$\cos^2 u = \frac{1+\cos^2 u}{2} \qquad \text{Power Reduction Formulas}$$

$$\Rightarrow \cos\frac{\theta}{2} = \pm\sqrt{\frac{1+\cos\theta}{2}}$$

We also have that:

In quadrant I, and quadrant IV, $\cos\frac{\theta}{2} > 0$

In quadrant II and quadrant III, $\cos\frac{\theta}{2} < 0$

Half-angle Formula for Tangent

In trigonometry, tangent half-angle formulas relate the tangent of half of an angle to trigonometric functions of the entire angle. Among these are the following:

$$\tan\left(\frac{\eta\pm\theta}{2}\right) = \frac{\sin\eta\pm\sin\theta}{\cos\eta+\cos\theta} = -\frac{\cos\eta-\cos\theta}{\sin\eta\mp\sin\theta},$$

$$\tan\left(\pm\frac{\theta}{2}\right) = \frac{\pm\sin\theta}{1+\cos\theta} = \frac{\pm\tan\theta}{\sec\theta+1} = \frac{\pm1}{\csc\theta+\cot\theta}, \qquad (\eta=0)$$

$$\tan\left(\pm\frac{\theta}{2}\right) = \frac{1-\cos\theta}{\pm\sin\theta} = \frac{\sec\theta-1}{\pm\tan\theta} = \pm(\csc\theta-\cot\theta), \quad (\eta=0)$$

$$\tan\left(\frac{1}{2}(\theta\pm\frac{\pi}{2})\right) = \frac{1\pm\sin\theta}{\cos\theta} = \sec\theta\pm\tan\theta = \frac{\csc\theta\pm1}{\cot\theta}, \qquad (\eta=\frac{\pi}{2})$$

$$\tan\left(\frac{1}{2}(\theta\pm\frac{\pi}{2})\right) = \frac{\cos\theta}{1\mp\sin\theta} = \frac{1}{\sec\theta\mp\tan\theta} = \frac{\cot\theta}{\csc\theta\mp1}, \qquad (\eta=\frac{\pi}{2})$$

$$\frac{1-\tan(\theta/2)}{1+\tan(\theta/2)} = \pm\sqrt{\frac{1-\sin\theta}{1+\sin\theta}}$$

$$\tan\frac{\theta}{2} = \pm\sqrt{\frac{1-\cos\theta}{1+\cos\theta}}$$

From these one can derive identities expressing the sine, cosine, and tangent as functions of tangents of half-angles:

$$\sin\alpha = \frac{2\tan\frac{\alpha}{2}}{1+\tan^2\frac{\alpha}{2}}$$

$$\cos\alpha = \frac{1-\tan^2\frac{\alpha}{2}}{1+\tan^2\frac{\alpha}{2}}$$

$$\tan\alpha = \frac{2\tan\frac{\alpha}{2}}{1-\tan^2\frac{\alpha}{2}}$$

Algebraic Proofs

Use double-angle formulae and $\sin^2\alpha + \cos^2\alpha = 1$,

$$\sin\alpha = 2\sin\frac{\alpha}{2}\cos\frac{\alpha}{2} = \frac{2\sin\frac{\alpha}{2}\cos\frac{\alpha}{2}}{\cos^2\frac{\alpha}{2}+\sin^2\frac{\alpha}{2}} = \frac{2\frac{\sin\frac{\alpha}{2}}{\cos\frac{\alpha}{2}}\frac{\cos\frac{\alpha}{2}}{\cos\frac{\alpha}{2}}}{\frac{\cos^2\frac{\alpha}{2}}{\cos^2\frac{\alpha}{2}}+\frac{\sin^2\frac{\alpha}{2}}{\cos^2\frac{\alpha}{2}}} = \frac{2\tan\frac{\alpha}{2}}{1+\tan^2\frac{\alpha}{2}}$$

$$\cos\alpha = \cos^2\frac{\alpha}{2}-\sin^2\frac{\alpha}{2} = \frac{\cos^2\frac{\alpha}{2}-\sin^2\frac{\alpha}{2}}{\cos^2\frac{\alpha}{2}+\sin^2\frac{\alpha}{2}} = \frac{\frac{\cos^2\frac{\alpha}{2}}{\cos^2\frac{\alpha}{2}}-\frac{\sin^2\frac{\alpha}{2}}{\cos^2\frac{\alpha}{2}}}{\frac{\cos^2\frac{\alpha}{2}}{\cos^2\frac{\alpha}{2}}+\frac{\sin^2\frac{\alpha}{2}}{\cos^2\frac{\alpha}{2}}} = \frac{1-\tan^2\frac{\alpha}{2}}{1+\tan^2\frac{\alpha}{2}}$$

Taking the quotient of the formulae for sine and cosine yields:

$$\tan\alpha = \frac{2\tan\frac{\alpha}{2}}{1-\tan^2\frac{\alpha}{2}}$$

Combining the Pythagorean identity $\cos^2\alpha+\sin^2\alpha=1$ with the double-angle formula for the cosine, $\cos 2\alpha=\cos^2\alpha-\sin^2\alpha=1-2\sin^2\alpha=2\cos^2\alpha-1$,

rearranging, and taking the square roots yields:

$$|\sin\alpha|=\sqrt{\frac{1-\cos 2\alpha}{2}} \text{ and } |\cos\alpha|=\sqrt{\frac{1+\cos 2\alpha}{2}}$$

which, upon division gives

$$|\tan\alpha|=\frac{\sqrt{1-\cos 2\alpha}}{\sqrt{1+\cos 2\alpha}}=\frac{\sqrt{1-\cos 2\alpha}\sqrt{1+\cos 2\alpha}}{1+\cos 2\alpha}=\frac{\sqrt{1-\cos^2 2\alpha}}{1+\cos 2\alpha}=\frac{|\sin 2\alpha|}{1+\cos 2\alpha}$$

or alternatively

$$|\tan\alpha|=\frac{\sqrt{1-\cos 2\alpha}}{\sqrt{1+\cos 2\alpha}}=\frac{1-\cos 2\alpha}{\sqrt{1+\cos 2\alpha}\sqrt{1-\cos 2\alpha}}=\frac{1-\cos 2\alpha}{\sqrt{1-\cos^2 2\alpha}}=\frac{1-\cos 2\alpha}{|\sin 2\alpha|}.$$

Also, using the angle addition and subtraction formulae for both the sine and cosine one obtains:

$$\cos(a+b)=\cos a\cos b-\sin a\sin b.$$
$$\cos(a-b)=\cos a\cos b+\sin a\sin b.$$
$$\sin(a+b)=\sin a\cos b+\cos a\sin b.$$
$$\sin(a-b)=\sin a\cos b-\cos a\sin b.$$

Pairwise addition of the above four formulae yields:

$$\begin{aligned}\sin(a+b)+\sin(a-b)&=\sin a\cos b+\cos a\sin b+\sin a\cos b-\cos a\sin b\\&=2\sin a\cos b\\\cos(a+b)+\cos(a-b)&=\cos a\cos b-\sin a\sin b+\cos a\cos b+\sin a\sin b\\&=2\cos a\cos b\end{aligned}$$

Setting $a = \dfrac{p+q}{2}$ and $b = \dfrac{p-q}{2}$ and substituting yields:

$$\sin\left(\frac{p+q}{2}+\frac{p-q}{2}\right)+\sin\left(\frac{p+q}{2}-\frac{p-q}{2}\right)=\sin(p)+\sin(q)$$

$$=2\sin\left(\frac{p+q}{2}\right)\cos\left(\frac{p-q}{2}\right)$$

$$\cos\left(\frac{p+q}{2}+\frac{p-q}{2}\right)+\cos\left(\frac{p+q}{2}-\frac{p-q}{2}\right)=\cos(p)+\cos(q)$$

$$=2\cos\left(\frac{p+q}{2}\right)\cos\left(\frac{p-q}{2}\right)$$

Dividing the sum of sines by the sum of cosines one arrives at:

$$\frac{\sin(p)+\sin(q)}{\cos(p)+\cos(q)}=\frac{2\sin\left(\frac{p+q}{2}\right)\cos\left(\frac{p-q}{2}\right)}{2\cos\left(\frac{p+q}{2}\right)\cos\left(\frac{p-q}{2}\right)}=\tan\left(\frac{p+q}{2}\right)$$

Geometric Proofs

Applying the formulae derived above to the rhombus figure on the right, it is readily shown that:

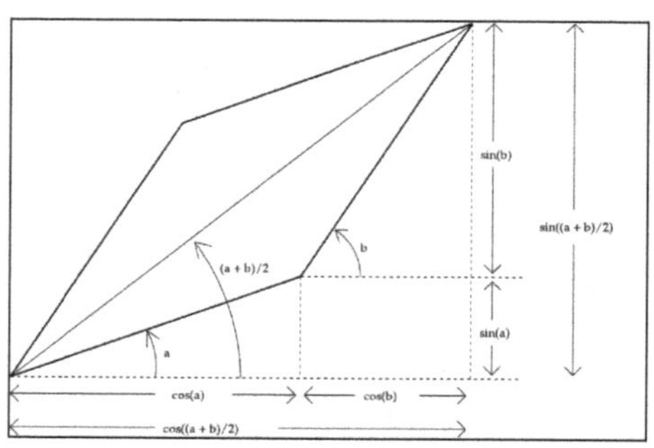

The sides of this rhombus have length 1. The angle between the horizontal line and the shown diagonal is (a + b)/2. This is a geometric way to prove a tangent half-angle formula. Note that sin((a + b)/2) and cos((a + b)/2) just show their relation to the diagonal, not the real value.

$$\tan\frac{a+b}{2}=\frac{\sin\frac{a+b}{2}}{\cos\frac{a+b}{2}}=\frac{\sin a+\sin b}{\cos a+\cos b}.$$

In the unit circle, application of the above shows that $t=\tan\left(\frac{\varphi}{2}\right)$. According to similar triangles,

$\frac{t}{\sin\varphi}=\frac{1}{1+\cos\varphi}$. It follows that

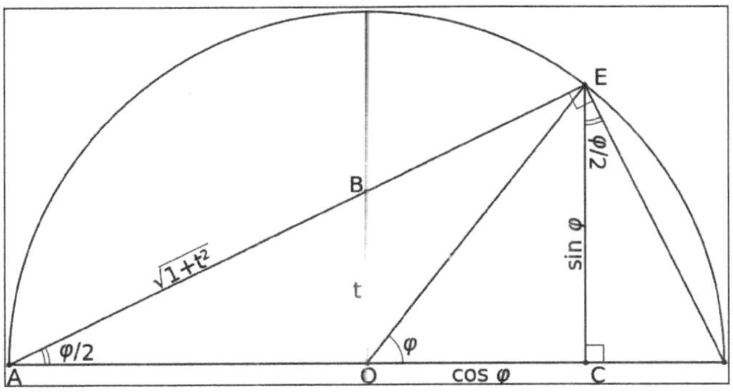

A geometric proof of the tangent half-angle formula:

$$t=\frac{\sin\varphi}{1+\cos\varphi}=\frac{\sin\varphi(1-\cos\varphi)}{(1+\cos\varphi)(1-\cos\varphi)}=\frac{1-\cos\varphi}{\sin\varphi}.$$

The Tangent Half-angle Substitution in Integral Calculus

In various applications of trigonometry, it is useful to rewrite the trigonometric functions (such as sine and cosine) in terms of rational functions of a new variable t. These identities are known collectively as the tangent half-angle formulae because of the definition of t. These identities can be useful in calculus for converting rational functions in sine and cosine to functions of t in order to find their antiderivatives.

Technically, the existence of the tangent half-angle formulae stems from the fact that the circle is an algebraic curve of genus 0. One then expects that the *circular functions* should be reducible to rational functions.

Geometrically, the construction goes like this: for any point (cos φ, sin φ) on the unit circle, draw the line passing through it and the point (−1, 0). This point crosses the y-axis at some point $y = t$. One can show using simple geometry that $t = \tan(\varphi/2)$. The equation for the drawn line is $y = (1 + x)t$. The equation for the intersection of the line and circle is then a quadratic equation involving t. The two solutions to this equation are (−1, 0) and ($\cos\varphi$, $\sin\varphi$). This allows us to write the latter as rational functions of t (solutions are given below).

Note also that the parameter t represents the stereographic projection of the point ($\cos\varphi$, $\sin\varphi$) onto the y-axis with the center of projection at (−1, 0). Thus, the tangent half-angle formulae give conversions between the stereographic coordinate t on the unit circle and the standard angular coordinate φ.

Then we have:

$$\cos\varphi = \frac{1-t^2}{1+t^2}, \quad \sin\varphi = \frac{2t}{1+t^2},$$
$$\tan\varphi = \frac{2t}{1-t^2} \quad \cot\varphi = \frac{1-t^2}{2t},$$
$$\sec\varphi = \frac{1+t^2}{1-t^2}, \quad \csc\varphi = \frac{1+t^2}{2t},$$

And,

$$e^{i\varphi} = \frac{1+it}{1-it}, \qquad e^{-i\varphi} = \frac{1-it}{1+it}.$$

By eliminating phi between the directly above and the initial definition of t, one arrives at the following useful relationship for the arctangent in terms of the natural logarithm:

$$\arctan t = \frac{1}{2i}\ln\frac{1+it}{1-it}.$$

In calculus, the Weierstrass substitution is used to find antiderivatives of rational functions of sin φ and cos φ. After setting:

$$t = \tan\tfrac{1}{2}\varphi.$$

This implies that:

$$\varphi = 2\arctan(t) + 2\pi n,$$

for some integer n, and therefore:

$$d\varphi = \frac{2\,dt}{1+t^2}.$$

Hyperbolic Identities

One can play an entirely analogous game with the hyperbolic functions. A point on (the right branch of) a hyperbola is given by (cosh θ, sinh θ). Projecting this onto y-axis from the center (−1, 0) gives the following:

$$t = \tanh\tfrac{1}{2}\theta = \frac{\sinh\theta}{\cosh\theta+1} = \frac{\cosh\theta-1}{\sinh\theta}$$

with the identities:

$$\cosh\theta = \frac{1+t^2}{1-t^2}, \quad \sinh\theta = \frac{2t}{1-t^2},$$

$$\tanh\theta = \frac{2t}{1+t^2}, \quad \coth\theta = \frac{1+t^2}{2t},$$

$$\operatorname{sech}\theta = \frac{1-t^2}{1+t^2}, \quad \operatorname{csch}\theta = \frac{1-t^2}{2t},$$

And,

$$e^{\theta} = \frac{1+t}{1-t}, \qquad e^{-\theta} = \frac{1-t}{1+t}.$$

The use of this substitution for finding antiderivatives was introduced by Karl Weierstrass.

Finding θ in terms of t leads to following relationship between the hyperbolic arctangent and the natural logarithm:

$$\operatorname{artanh} t = \frac{1}{2}\ln\frac{1+t}{1-t}.$$

("ar-" is used rather than "arc-" because "arc" is about arc length and "ar" abbreviates "area". It is the area between two rays and a hyperbola, rather than the arc length between two rays measure along an arc a circle).

The Gudermannian Function

Comparing the hyperbolic identities to the circular ones, one notices that they involve the same functions of t, just permuted. If we identify the parameter t in both cases we arrive at a relationship between the circular functions and the hyperbolic ones. That is, if:

$$t = \tan\tfrac{1}{2}\varphi = \tanh\tfrac{1}{2}\theta$$

Then,

$$\varphi = 2\tan^{-1}\tanh\tfrac{1}{2}\theta \equiv \operatorname{gd}\theta.$$

where $\operatorname{gd}(\theta)$ is the Gudermannian function. The Gudermannian function gives a direct relationship between the circular functions and the hyperbolic ones that does not involve complex numbers. The above descriptions of the tangent half-angle formulae (projection the unit circle and standard hyperbola onto the y-axis) give a geometric interpretation of this function.

Pythagorean Triples

The tangent of half of an acute angle of a right triangle whose sides are a Pythagorean triple will necessarily be a rational number in the interval (0, 1). Vice versa, when a half-angle tangent is a rational number in the interval (0, 1), there is a right triangle that has the full angle and that has side lengths that are a Pythagorean triple.

Triple-angle Formulae

The trigonometric triple-angle identities give a relationship between the basic trigonometric functions applied to three times an angle in terms of trigonometric functions of the angle itself.

Theorem

Triple-angle Identities

$$\sin 3\theta = 3\sin\theta - 4\sin^3\theta$$

$$\cos 3\theta = 4\cos^3\theta - 3\cos\theta$$

Proof:

To prove the triple-angle identities, we can write sin 3θ as $\sin(2\theta+\theta)$. Then we can use the sum formula and the double-angle identities to get the desired form:

$$\begin{aligned}\sin 3\theta &= \sin(2\theta+\theta)\\ &= \sin 2\theta\cos\theta + \cos 2\theta\sin\theta\\ &= (2\sin\theta\cos\theta)\cos\theta + (1-2\sin^2\theta)\sin\theta\\ &= 2\sin\theta\cos^2\theta + \sin\theta - 2\sin^3\theta\\ &= 2\sin\theta(1-\sin^2\theta) + \sin\theta - 2\sin^3\theta\\ &= 2\sin\theta - 2\sin^3\theta + \sin\theta - 2\sin^3\theta\\ &= 3\sin\theta - 4\sin^3\theta.\end{aligned}$$

The triple angle identity of cos 3θ can be proved in a very similar manner.

From these formulas, we also have the following identities for $\sin^3(\theta)$ and $\cos^3(\theta)$ in terms of lower powers:

$$\sin^3(\theta) = \frac{3\sin(\theta)-\sin(3\theta)}{4}, \cos^3(\theta) = \frac{\cos(3\theta)+3\cos(\theta)}{4}.$$

To remember the cosine formula, the trick that I like to use is to read cosine as "dollar."

Then, we say:

"Dollar thirty is equal to four dollar thirty minus three dollar".

$$\begin{aligned} &\$1.30 &&= \$4.30 &&- \$3 \\ &1\cos 3\theta &&= 4\cos^3\theta &&-3\cos\theta \end{aligned}$$

Just leaving a mark here,

$$\begin{aligned} \tan x \tan(60° - x)\tan(60° + x) &= \quad \tan(3x) \\ 4\sin x \sin(60° - x)\sin(60° + x) &= \quad \sin(3x) \\ 4\cos x \cos(60° - x)\cos(60° + x) &= \quad \cos(3x) \end{aligned}$$

$$\tan(3x) = \frac{3\tan x - \tan^3 x}{1 - 3\tan^2 x}.$$

Example:

Prove that:

$$\tan(6°) = \tan(12°)\tan(24°)\tan(48°).$$

Triple-angle Formula for Sine

Theorem:

$$\sin 3\theta = 3\sin\theta - 4\sin^3\theta$$

where sin denotes sine.

Proof:

$$\sin 3 = \sin(2\theta + \theta)$$

$= \sin 2\theta \cos\theta + \cos 2\theta \sin\theta$ Sine of Sum

$= (2\sin\theta\cos\theta)\cos\theta + (\cos^2\theta - \sin^2\theta)\sin\theta$ Double Angle Formula for Sine and Double Angle Formula for Cosine

$= 2\sin\theta\cos^2\theta + \cos^2\theta\sin\theta - \sin^3\theta$

$= 2\sin\theta(1 - \sin^2\theta) + (1 - \sin^2\theta)\sin\theta - \sin^3\theta$ Sum of Squares of Sine and Cosine

$= 2\sin\theta - 2\sin^3\theta + \sin\theta - \sin^3\theta - \sin^3\theta$ multiplying out

$= 3\sin\theta - 4\sin^3\theta$ gathering terms

Proof:

We have:

$$\cos 3\theta + i\sin 3\theta = (\cos\theta + i\sin\theta)^3$$

$$= (\cos\theta)^3 + \binom{3}{1}(\cos\theta)^2(i\sin\theta) \qquad \text{De Moivre's Formula}$$

$$+\binom{3}{2}(\cos\theta)(i\sin\theta)^2 + (i\sin\theta)^3 \qquad \text{Binomial Theorem}$$

$= \cos^3\theta + 3i\cos^2\theta\sin\theta + 3i^2\cos\theta\sin^2\theta + i^3\sin^3\theta$ substituting for binomial coefficients

$= \cos^3\theta + 3i\cos^2\theta\sin\theta - 3\cos\theta\sin^2\theta - i\sin^3\theta \qquad i^2 = -1$

$= \cos^3\theta - 3\cos\theta\sin^2\theta$

$+i\left(3\cos^2\theta\sin\theta - \sin^3\theta\right)$ rearranging

Hence:

$$\begin{aligned}
\sin 3\theta &= 3\cos^2\theta\sin\theta - \sin^3\theta && \text{equating imaginary parts in (1)}\\
&= 3(1-\sin^2\theta)\sin\theta - \sin^3\theta && \text{Sum of Squares of Sine and Cosine}\\
&= 3\sin\theta - 4\sin^3\theta && \text{multiplying out and gathering terms}
\end{aligned}$$

Triple-angle Formula for Cosine

Theorem:

$$\cos 3\theta = 4\cos^3\theta - 3\cos\theta$$

where cos denotes cosine.

Example: 2 cos 3θ + 1

$$2\cos 3\theta + 1 = \left(\cos\theta - \cos\frac{2\pi}{9}\right)\left(\cos\theta - \cos\frac{4\pi}{9}\right)\left(\cos\theta - \cos\frac{8\pi}{9}\right)$$

Proof:

$$\cos 3\theta = \cos(2\theta + \theta)$$

$= \cos 2\theta \cos\theta - \sin 2\theta \sin\theta$ Cosine of Sum

$= (\cos^2\theta - \sin^2\theta)\cos\theta - (2\sin\theta\cos\theta)\sin\theta$ Double Angle Formula for Cosine and Double Angle Formula for Sine

$= \cos^3\theta - \sin^2\theta\cos\theta - 2\sin^2\theta\cos\theta$

$= \cos^3\theta - \sin^2\theta\cos\theta - 2\sin^2\theta\cos\theta$ Sum of Squares of Sine and Cosine

$= \cos^3\theta - \cos\theta + \cos^3\theta - 2\cos\theta + 2\cos^3\theta$ multiplying out

$= 4\cos^3\theta - 3\cos\theta$ gathering terms

Proof:

We have:

$\cos 3\theta + i\sin 3\theta = (\cos\theta + i\sin\theta)^3$ De Moivre's Formula

$= (\cos\theta)^3 + \binom{3}{1}(\cos\theta)^2(i\sin\theta)$ Binomial Theorem

$+\binom{3}{2}(\cos\theta)(i\sin\theta)^2 + (i\sin\theta)^3$

$= \cos^3\theta + 3i\cos^2\theta\sin\theta + 3i^2\cos\theta\sin^2\theta + i^3\sin^3\theta$ substituting for binomial coefficients

$= \cos^3\theta + 3i\cos^2\theta\sin\theta - 3\cos\theta\sin^2\theta - i\sin^3\theta$

$= \cos^3\theta - 3\cos\theta\sin^2\theta$ $i^2 = -1$

$+i(3\cos^2\theta\sin\theta - \sin^3\theta)$ rearranging

Hence:

$\cos 3\theta = \cos^3\theta - 3\cos\theta\sin^2\theta$ equating real parts in (1)

$= \cos^3\theta - 3\cos\theta(1 - \cos^2\theta)$ Sum of Squares of Sine and Cosine

$= 4\cos^3\theta - 3\cos\theta$ multiplying out and gathering terms

Proof:

$$\cos 3\theta = \cos(2\theta + \theta)$$

$= \cos 2\theta \cos\theta - \sin 2\theta \sin\theta$ Cosine of Sum

$= (2\cos^2\theta - 1)\cos\theta - 2\sin^2\theta\cos\theta$ Double Angle Formula for Cosine and Double Angle Formula for Sine

$= (2\cos^2\theta - 1 - 2(1 - 2\cos^2\theta))\cos\theta$ Sum of Squares of Sine and Cosine

$= 4\cos^3\theta - 3\cos\theta$ gathering terms

Triple-angle Formula for Tangent

Theorem:

$$\tan(3\theta) = \frac{3\tan\theta - \tan^3\theta}{1 - 3\tan^2\theta}$$

where tan denotes tangent.

Proof:

$\tan(3\theta) = \frac{\sin(3\theta)}{\cos(3\theta)}$ Tangent is Sine divided by Cosine

$= \frac{3\sin\theta - 4\sin^3\theta}{4\cos^3\theta - 3\cos\theta}$ Triple Angle Formula for Sine and Triple Angle Formula for Cosine

$$= \frac{3\sin\theta - 4\sin^3\theta}{4\cos^3\theta - 3\cos\theta}\frac{\cos^3\theta}{\cos^3\theta}$$

$= \frac{\frac{3\tan}{\cos^2\theta} - 4\tan^3\theta}{4 - \frac{3\cos\theta}{\cos^2\theta}}$ Tangent is Sine divided by Cosine

$= \frac{3\tan\theta\sec^2\theta - 4\tan^3\theta}{4 - 3\sec^2\theta}$ Secant is Reciprocal of Cosine

$= \frac{3\tan\theta(1 + \tan^2\theta) - 4\tan^3\theta}{4 - 3(1 + \tan^2\theta)}$ Difference of Squares of Secant and Tangent

$$= \frac{3\tan\theta + 3\tan^3\theta - 4\tan^3\theta}{4 - 3 - 3\tan^2\theta} \quad \text{multiplying out}$$

$$= \frac{3\tan\theta - \tan^3\theta}{1 - 3\tan^2\theta} \qquad \text{gathering terms}$$

Proof:

$$\tan(3\theta) = \frac{\tan\theta + \tan(2\theta)}{1 - \tan\theta\tan(2\theta)} \quad \text{Tangent of Sum}$$

$$= \frac{\tan\theta + \dfrac{2\tan\theta}{1-\tan^2\theta}}{1 - \tan\theta\dfrac{2\tan\theta}{1-\tan^2\theta}} \quad \text{Double Angle Formula for Tangent}$$

$$= \frac{\tan\theta\left(1-\tan^2\theta\right) + 2\tan\theta}{\left(1-\tan^2\theta\right) - 2\tan^2\theta} \quad \text{simplifying}$$

$$= \frac{\tan\theta - \tan^3\theta + 2\tan\theta}{1 - \tan^2\theta - 2\tan^2\theta} \quad \text{simplifying}$$

$$= \frac{3\tan\theta - \tan^3\theta}{1 - 3\tan^2\theta} \qquad \text{simplifying}$$

Law of Sines

Theorem

For any triangle ΔABC:

$$\frac{a}{sinA} = \frac{b}{sinB} = \frac{c}{sinC}$$

where $a, b,$ and c are the sides opposite A, B and C respectively.

Proof:

Construct the altitude from B.

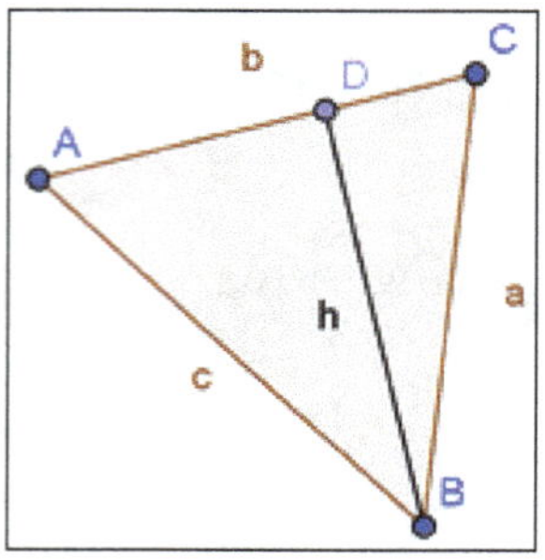

It can be seen from the definition of sine that:

$$\sin A = \frac{h}{c} \text{ and } sinC = \frac{h}{a}$$

Thus,

$$h = c\sin A \text{ and } h = a\,sinC$$

This gives:

$$c\,sinA = a\,sinC$$

So,

$$\frac{a}{sinA} = \frac{c}{sinC}$$

Similarly, constructing the altitude from A gives:

$$\frac{b}{sinB} = \frac{c}{sinC}$$

Proof:

Construct the circumcircle of ΔABC, let O be the circumcenter and R be the circumradius.

Construct ΔAOB and let E be the foot of the altitude of ΔAOB from O.

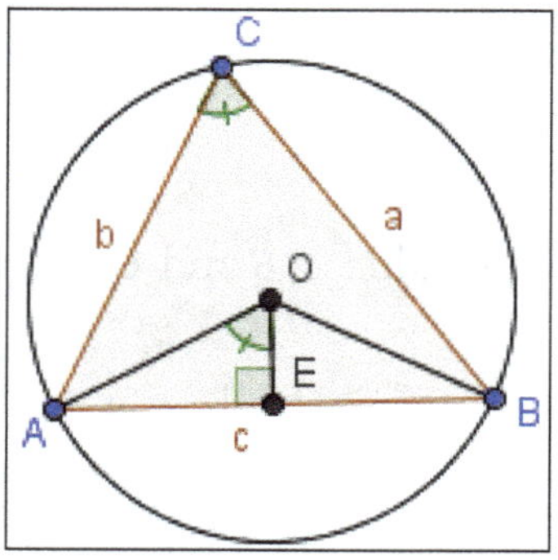

By the Inscribed Angle Theorem:

$$\angle ACB = \frac{\angle AOB}{2}$$

From the definition of the circumcenter:

$$AO = BO$$

From the definition of altitude and the fact that all right angles are congruent:

$$\angle AEO = \angle BEO$$

Therefore from Pythagoras's Theorem:

$$AE = BE$$

and then from Triangle Side-Side-Side Equality:

$$\angle AOE = \angle BOE$$

Thus,

$$\angle ACB = \angle AOE$$

Then by the definition of sine:

$$\sin C = \sin(\angle AOE) = \frac{c/2}{R}$$

and so:

$$\frac{c}{sinC} = 2R$$

Because the same argument holds for all three angles in the triangle:

$$\frac{c}{sinC} = 2R = \frac{b}{sinB} = 2R = \frac{a}{sinA}$$

Note that this proof also yields a useful extension of the law of sines:

$$\frac{a}{sinA} = \frac{b}{sinB} = \frac{c}{sinC} = 2R$$

This result is also known as the sine law, sine rule or rule of sines.

Law of Cosines

Theorem:

Let ΔABC be a triangle whose sides a, b, c are such that a is opposite A, b is opposite B and c is opposite C.

Then,

$$c^2 = a^2 + b^2 - 2ab \cos C$$

Proof:

Let ΔABC be embedded in a Cartesian coordinate system by identifying:

$$C := (0,0)$$
$$B := (a,0)$$

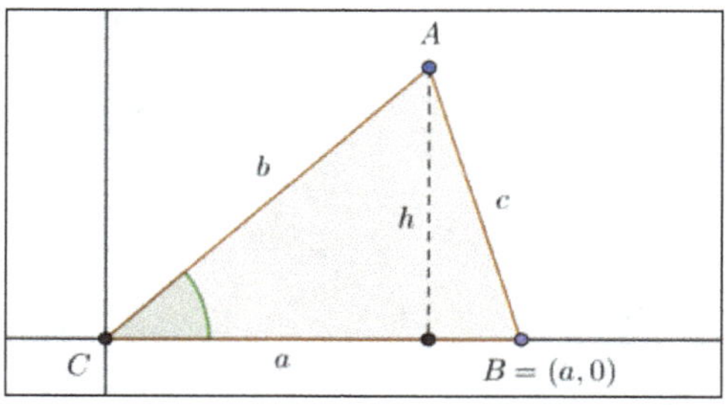

Thus by definition of sine and cosine:

$$A = (b\cos C, b\sin C)$$

By the Distance Formula:

$$c = \sqrt{(b\cos C - a)^2 + (b\sin C - 0)^2}$$

Hence:

$c^2 (b\cos C - a)^2 + (b\sin C - 0)^2$	squaring both sides of Distance Formula
$= b^2\cos^2 C - 2ab\cos C + a^2 + b^2\sin^2 C$	Square of Difference
$= a^2 + b^2\left(\sin^2 C + \cos^2 C\right) - 2ab\cos C$	Real Multiplication Distributes over Addition
$= a^2 + b^2 - 2ab\cos C$	Sum of Squares of Sine and Cosine

Proof:

Let ΔABC be a triangle.

Case 1: AC greater than AB

Using AC as the radius, we construct a circle whose center is A.

Now we extend:

CB to D
AB to F
BA to G
CA to E.

D is joined with E, thus:

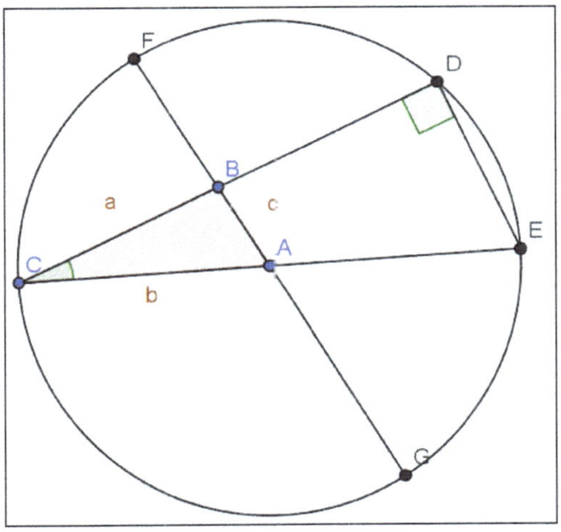

Using the Intersecting Chord Theorem we have:

$$GB \cdot BF = CB \cdot BD$$

AF is a radius, so $AF = AC = b = GA$ and thus:

$$GB = GA + AB = b + c$$
$$BF = AF - AB = b - c$$

Thus:

$$(b+c)(b-c) = a \cdot BD$$

$$\Rightarrow \quad \frac{b^2 - c^2}{a} = BD$$

Next:

$$
\begin{aligned}
CD &= CB + BD \\
&= a + \frac{b^2 - c^2}{a} \\
&= \frac{a^2 + b^2 - c^2}{a}
\end{aligned}
$$

As CA is a radius, CE is a diameter.

By Thales' Theorem, it follows that $\angle CDE$ is a right angle.

Then using the definition of cosine, we have:

$$
\begin{aligned}
\cos C &= \frac{CD}{CE} \\
&= \frac{\left(\dfrac{a^2 + b^2 - c^2}{a}\right)}{2b} \\
&= \frac{a^2 + b^2 - c^2}{2ab} \\
\Rightarrow \quad c^2 &= a^2 + b^2 - 2ab\cos C
\end{aligned}
$$

Case 2: AC less than AB

When AC is less than AB, the point B lies outside the circle and so the diagram needs to be modified accordingly.

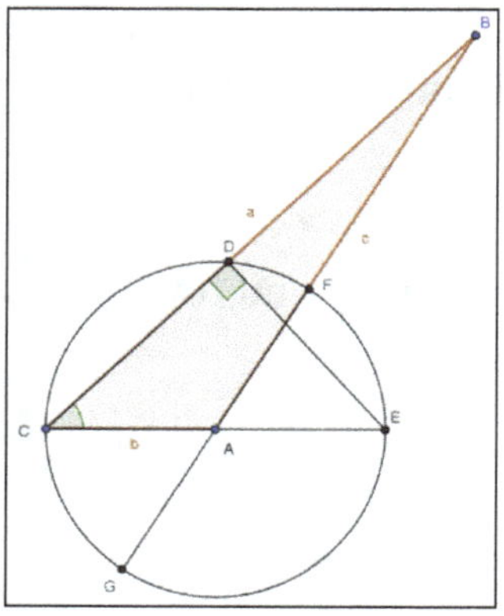

Now we extend:

BA to G

CA to E.

Then we construct:

D as the point at which CB intersects the circle.

F as the point at which AB intersects the circle.

Finally D is joined to E.

Using the Secant Secant Theorem we have:

$$GB \cdot BF = CB \cdot BD$$

AF is a radius, so $AF = AC = b = GA$ and thus:

$$GB = GA + AB = b + c$$
$$BF = AB - AF = b - c$$

Thus:

$$(b+c)(b-c) = CB \cdot BD \qquad \text{Secant Secant Theorem}$$
$$(b+c)(b-c) = a \cdot BD$$
$$\Rightarrow \qquad \frac{b^2 - c^2}{a} = BD$$

Next:

$$\begin{aligned} CD &= CB - BD \\ &= a - \frac{b^2 - c^2}{a} \\ &= \frac{a^2 - b^2 + c^2}{a} \end{aligned}$$

As CA is a radius, CE is a diameter.

By Thales' Theorem, it follows that $\angle CDE$ is a right angle.

Then using the definition of cosine, we have:

$$\begin{aligned} \cos C &= \frac{CD}{CE} \\ &= \frac{\left(\frac{a^2 - b^2 + c^2}{a}\right)}{2b} \\ &= \frac{a^2 - b^2 + c^2}{2ab} \end{aligned}$$

$$\Rightarrow \qquad c^2 = a^2 + b^2 - 2ab\cos C$$

Case 3: $AC = AB$

When $AC = AB$ the points B, D and F coincide on the circumference of the circle:

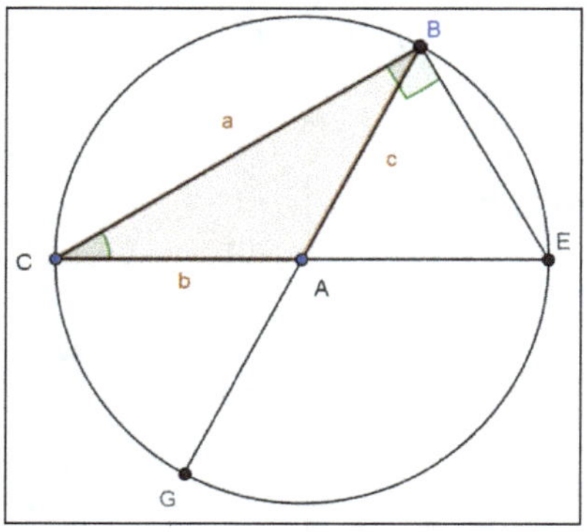

We extend:

BA to G

CA to E

and immediately:

$GB = CB$

Proof:

Lemma: Right Triangle

Let ΔABC be a right triangle such that $\angle A$ is right.

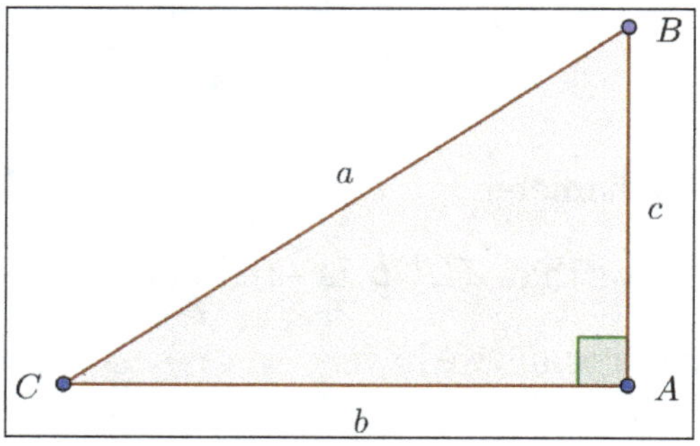

$a^2 = b^2 + c^2$	Pythagoras's Theorem
$c^2 = a^2 - b^2$	adding $-b^2$ to both sides and rearranging
$= a^2 - 2b^2 + b^2$	adding $0 = b^2 - b^2$ to the right hand side
$= a^2 2ab\left(\frac{b}{a}\right) + b^2$	multiplying $2b^2$ by $\frac{a}{a}$
$= a^2 + b^2 - 2ab\cos C$	Definition of Cosine : $C = \frac{b}{a}$

Hence the result.

Acute Triangle

Let ΔABC be an acute triangle.

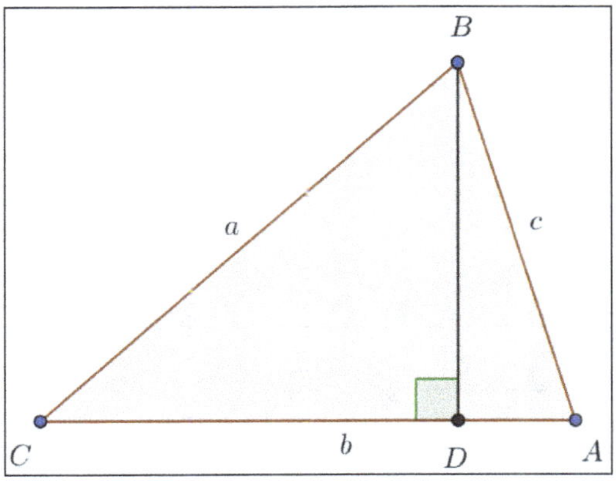

Let BD be dropped perpendicular to AC and let us define $h = BD, e = CD$ and $f = AD$.

Then ΔCDB and ΔADB are right triangles.

So we have both:

$$b^2 = (e+f)^2 = e^2 + f^2 + 2ef$$

$$e = a\cos C \text{ Definition:Cosine of Angle}$$

We'll start with the first equation and use the rest of them to get the desired result:

$$\begin{aligned}
c^2 &= h^2 + f^2 \\
&= a^2 - e^2 + f^2 \\
&= a^2 - e^2 + f^2 + 2e^2 - 2e^2 + 2ef - 2ef && \text{adding and subtracting } 2e^2 \text{ and } 2ef \\
&= a^2 + (e^2 + f^2 + 2ef) - 2e(e+f) && \text{rearanging} \\
&= a^2 + b^2 - 2ab\cos C && \text{using to substitute both parentheses for } b^2 \text{ and } b \text{ respectively, and to subst. e for } a\cos C
\end{aligned}$$

Obtuse Triangle

Let ΔABC be an obtuse triangle.

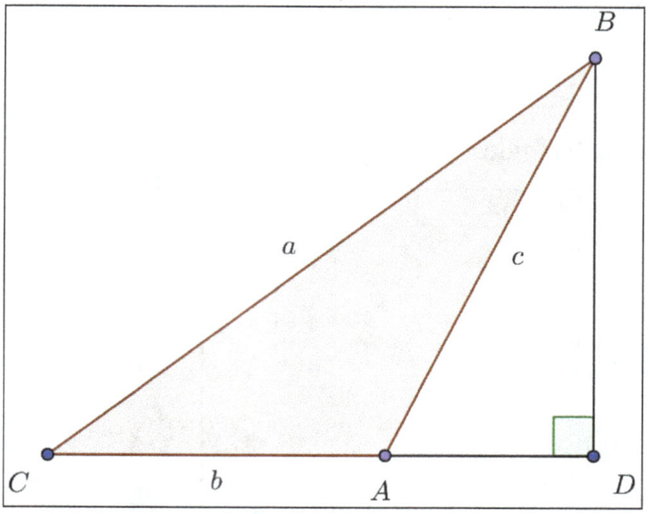

Let AC be extended and BD be dropped perpendicular to AC, and let us define $h = BD, e = CD$ and $f = AD$.

Then ΔCDB and $\Delta\ ADB$ are right triangles.

So we have both:

$c^2 = h^2 + f^2$ Pythagoras's Theorem

$a^2 = h^2 + e^2$ Pythagoras's Theorem

and also:

$e^2 = (b+f)^2 = b^2 + f^2 + 2bf$

$e = a\cos C$ Definition:Cosine of Angle

We'll start with the first equation and use the rest of them to get the desired result:

$c^2 = h^2 + f^2$

$= a^2 - e^2 + f^2$

$= a^2 - b^2 - f^2 - 2bf + f^2$

$= a^2 - b^2 - 2bf + 2b^2 - 2b^2$ canceling out $f^2 - f^2$ and adding and subtracting $2b^2$

$= a^2 + b^2 - 2b(b+f)$ rearanging

$= a^2 + b^2 - 2ab\cos C$ using to substitute $b + f = e$ with $a\cos C$

Law of Tangents

Theorem: Let $\triangle ABC$ be a triangle whose sides a, b, c are such that a is opposite A, b is opposite B and c is opposite C.

Then:

$$\frac{a+b}{a-b}=\frac{\tan\frac{1}{2}(A+B)}{\tan\frac{1}{2}(A-B)}$$

Proof:

Let $d=\frac{a}{\sin A}$.

From the Law of Sines, let:

$$d=\frac{a}{\sin A}=\frac{b}{\sin B}$$

so that:

$$\begin{aligned} a &= d\sin A \\ b &= d\sin B \\ \Rightarrow \frac{a+b}{a-b} &= \frac{d\sin A+d\sin B}{d\sin A-d\sin B} \\ &= \frac{\sin A+\sin B}{\sin A-\sin B} \end{aligned}$$

Let $C=\frac{1}{2}(A+B)$ and $D=\frac{1}{2}(A-B)$, and proceed as follows:

$$\Rightarrow \frac{a+b}{a-b}=\frac{2\sin C\cos D}{\sin A-\sin B} \quad \text{Prosthaphaeresis Formula for Sine plus Sine}$$

$$=\frac{2\sin C\cos D}{2\sin D\cos C} \quad \text{Prosthaphaeresis Formula for Sine minus Sine}$$

$$=\frac{\frac{\sin C}{\cos C}}{\frac{\sin D}{\cos D}} \quad \text{dividing top and bottom by } \cos C\cos D$$

$$= \frac{\tan C}{\tan D} \qquad \text{Tangent is Sine divided by Cosine}$$

$$= \frac{\tan \frac{1}{2}(A+B)}{\tan \frac{1}{2}(A-B)} \qquad \text{substituting back for } C \text{ and } D$$

Morrie's Law

Morrie's law is the trigonometric identity:

$$\cos(20^\circ) \cdot \cos(40^\circ) \cdot \cos(80^\circ) = \frac{1}{8}.$$

It is a special case of the more general identity:

$$2^n \cdot \prod_{k=0}^{n-1} \cos(2^k \alpha) = \frac{\sin(2^n \alpha)}{\sin(\alpha)}$$

with $n = 3$ and $\alpha = 20^\circ$ and the fact that:

$$\frac{\sin(160^\circ)}{\sin(20^\circ)} = \frac{\sin(180^\circ - 20^\circ)}{\sin(20^\circ)} = 1,$$

since:

$$\sin(180^\circ - x) = \sin(x).$$

The name is due to the physicist Richard Feynman, who used to refer to the identity under that name. Feynman picked that name because he learned it during his childhood from a boy with the name Morrie Jacobs and afterwards remembered it for all of his life.

A similar identity for the sine function also holds:

$$\sin(20^\circ) \cdot \sin(40^\circ) \cdot \sin(80^\circ) = \frac{\sqrt{3}}{8}.$$

Moreover, dividing the second identity by the first, the following identity is evident:

$$\tan(20^\circ) \cdot \tan(40^\circ) \cdot \tan(80^\circ) = \sqrt{3} = \tan(60^\circ).$$

Proof:

Recall the double angle formula for the sine function:

$$\sin(2\alpha) = 2\sin(\alpha)\cos(\alpha).$$

Solve for $\cos(\alpha)$

$$\cos(\alpha) = \frac{\sin(2\alpha)}{2\sin(\alpha)}.$$

It follows that:

$$\cos(2\alpha) = \frac{\sin(4\alpha)}{2\sin(2\alpha)}$$

$$\cos(4\alpha) = \frac{\sin(8\alpha)}{2\sin(4\alpha)}$$

$$\vdots$$

$$\cos(2^{n-1}\alpha) = \frac{\sin(2^n\alpha)}{2\sin(2^{n-1}\alpha)}.$$

Multiplying all of these expressions together yields:

$$\cos(\alpha)\cos(2\alpha)\cos(4\alpha)\cdots\cos(2^{n-1}\alpha) = \frac{\sin(2\alpha)}{2\sin(\alpha)}\cdot\frac{\sin(4\alpha)}{2\sin(2\alpha)}\cdot\frac{\sin(8\alpha)}{2\sin(4\alpha)}\cdots\frac{\sin(2^n\alpha)}{2\sin(2^{n-1}\alpha)}.$$

The intermediate numerators and denominators cancel leaving only the first denominator, a power of 2 and the final numerator. Note that there are n terms in both sides of the expression. Thus:

$$\prod_{k=0}^{n-1}\cos(2^k\alpha) = \frac{\sin(2^n\alpha)}{2^n\sin(\alpha)},$$

which is equivalent to the generalization of Morrie's law.

Euler's Formula

In complex analysis, Euler's formula provides a fundamental bridge between the exponential function and the trigonometric functions. For complex numbers x, Euler's formula says that:

$$e^{ix} = \cos x + i\sin x.$$

In addition to its role as a fundamental mathematical result, Euler's formula has numerous applications in physics and engineering.

Proof of Euler's Formula

A straightforward proof of Euler's formula can be had simply by equating the power series representations of the terms in the formula:

$$\cos x = 1 - \frac{x^2}{2!} + \frac{x^4}{4!} - \cdots$$

and,

$$\sin x = x - \frac{x^3}{3!} + \frac{x^5}{5!} - \cdots,$$

so,

$$\begin{aligned} \cos x + i \sin x &= 1 + ix - \frac{x^2}{2!} - i\frac{x^3}{3!} + \frac{x^4}{4!} - \cdots \\ &= 1 + ix + \frac{(ix)^2}{2!} + \frac{(ix)^3}{3!} + \frac{(ix)^4}{4!} + \cdots \\ &= e^{ix}. \end{aligned}$$

Theorem:

Euler's Formula

Suppose x is complex. Then:

$$e^{ix} = \cos x + i \sin x.$$

Example:

Compute $e^{i\pi}$.

We have:

$$e^{i\pi} = \cos \pi + i \sin \pi = -1,$$

which leads to the very famous Euler's identity: $e^{i\pi} + 1 - 0$.

Example:

Compute i^i.

Recall that $\forall k \in \mathbb{N}$ we have that:

$$e^{i(\pi/2+2\pi k)} = i$$
$$\Rightarrow \left(e^{i(\pi/2+2\pi k)}\right)^i = i^i.$$

Therefore,

$$i^i = e^{i^2(\pi/2+2\pi k)} = e^{-\pi/2-2\pi k}.$$

Note: This means that i^i is not a well defined (unique) quantity. To remedy this, one needs to specify a branch cut. For example, we can define the argument of $e^{i\theta}$ to be defined for $\theta \in [0, 2\pi)$, in which case we have that $i^i = e^{-\pi/2}$. That is, this forces $k = 0$. Of course, different branch cut can be chosen yielding different values for k.

Geometric Interpretation

Euler's formula allows for any complex number x to be represented as e^{ix}, which sits on a unit circle with real and imaginary components $\cos x$ and $\sin x$, respectively. Various operations (such as finding the roots of unity) can then be viewed as rotations along the unit circle.

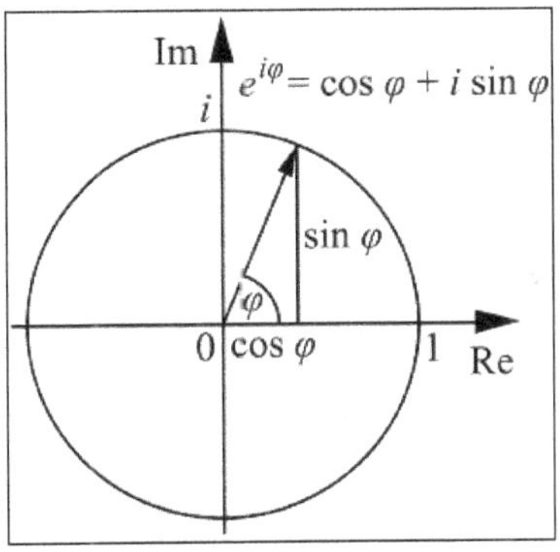

Trigonometric Applications

One immediate application of Euler's formula is to extend the definition of the trigonometric functions to allow for arguments that extend the range of the functions beyond what is allowed under the real numbers.

A couple useful results to have at hand are the facts that:

$$e^{-ix} = \cos x - i\sin x,$$

so,

$$e^{ix} + e^{-ix} = 2\cos x.$$

It follows that:

$$\cos x = \frac{e^{ix} + e^{-ix}}{2}, \qquad \text{and similarly}$$

$$\sin x = \frac{e^{ix} - e^{-ix}}{2i}$$

and,

$$\tan x = \frac{e^{ix} - e^{-ix}}{i\left(e^{ix} + e^{-ix}\right)}.$$

Example:

Solve $\cos x = 2$ in the complex numbers.

We first note that if $x = x_0$ is a solution, then so is $x = 2\pi k \pm x_0$ for any integer k. This is because $\cos x$ is an even function with a fundamental period of 2π. Taking $\cos x = \frac{e^{ix} + e^{-ix}}{2}$ yields:

$$e^{ix} + e^{-ix} = 4$$

$$\left(e^{ix}\right)^2 - 4e^{ix} + 1 = 0$$

$$e^{ix} = 2 \pm \sqrt{3}$$

$$\Rightarrow x = \frac{1}{i} \text{ In } \left(2 \pm \sqrt{3}\right)$$

$$= -i \text{ In } \left(2 \pm \sqrt{3}\right).$$

Hence $x = 2\pi k \pm i \text{ In } \left(2 \pm \sqrt{3}\right), 2\pi k \mp i \text{ In}\left(2 \pm \sqrt{3}\right)$ for any integer k.

Euler's formula also allows for the derivation of several trigonometric identities quite easily. Starting with:

$$e^{i(x \pm y)} = \cos(x \pm y) + i \sin(x \pm y),$$

one finds:

$$e^{i(x \pm y)} = e^{ix} e^{\pm iy}$$

$$= (\cos x + i \sin x)(\cos y \pm i \sin y)$$

$$= \cos x \cos y \mp \sin x \sin y + i(\sin x \cos y \mp \cos x \sin y).$$

Equating the real and imaginary parts, respectively, yields the familiar sum and difference formulas:

$$\cos(x \pm y) = \cos x \cos y \mp \sin x \sin y$$

and,

$$\sin(x \pm y) = \sin x \cos y \pm \cos x \sin y.$$

De Moivre's Theorem

De Moivre's theorem gives a formula for computing powers of complex numbers. We first gain some intuition for de Moivre's theorem by considering what happens when we multiply a complex number by itself.

Recall that using the polar form, any complex number $z = a + ib$ can be represented as $z = r(\cos\theta + i\sin\theta)$ with:

$$\text{Absolute value}: r = \sqrt{a^2 + b^2}$$

$$\text{Argument } \theta \text{ subject to}: \cos\theta = \frac{a}{r}, \sin\theta = \frac{b}{r}.$$

Then squaring the complex number z gives:

$$\begin{aligned} z^2 &= (r(\cos\theta + i\sin\theta))^2 \\ &= r^2(\cos\theta + i\sin\theta)^2 \\ &= r^2(\cos\theta\cos\theta + i\sin\theta\cos\theta + i\sin\theta\cos\theta + i^2\sin\theta\sin\theta) \\ &= r^2((\cos\theta\cos\theta - \sin\theta\sin\theta) + i(\sin\theta\cos\theta + \sin\theta\cos\theta)) \\ &= r^2(\cos 2\theta + i\sin 2\theta). \end{aligned}$$

This shows that by squaring a complex number, the absolute value is squared and the argument is multiplied by 2. For $n \geq 3$, de Moivre's theorem generalizes this to show that to raise a complex number to the nth power, the absolute value is raised to the nth power and the argument is multiplied by n.

Theorem

For any complex number x and any integer n,

$$(r(\cos\theta + i\sin\theta))^n = r^n(\cos(n\theta) + i\sin(n\theta)).$$

Proof:

We'll prove by induction.

We have:

$$z^n = (r(\cos(\theta) + i\sin(\theta))^n$$
$$= r^n(\cos(\theta) + i\sin(\theta))^n.$$

Let's focus on the second part: $(\cos(\theta) + i\sin(\theta))^n$. For $n = 1$, we have:

$$(\cos(\theta) + i\sin(\theta))^1 = \cos(1 \cdot \theta) + i\sin(1 \cdot \theta),$$

which is true.

We can assume the same formula is true for $n = k$, so we have:

$$(\cos(\theta) + i\sin(\theta))^k = \cos(k\theta) + i\sin(k\theta).$$

For $n = k + 1$, we expect to have:

$$(\cos(\theta) + i\sin(\theta))^{k+1} = \cos((k+1)\theta) + i\sin((k+1)\theta).$$

We get:

$$(\cos(\theta) + i\sin(\theta))^{k+1}$$

$$= (\cos(\theta) + i\sin(\theta))^k (\cos(\theta) + i\sin(\theta))^1$$

$= (\cos(k\theta) + i\sin(k\theta))(\cos(1 \cdot \theta) + i\sin(1 \cdot \theta))$ (We assume this to be true for $x = k$.)

$= \cos(k\theta)\cos(\theta) + \cos(k\theta)i\sin(\theta) + i\sin(k\theta)\cos(\theta) + i^2\sin(k\theta)\sin(\theta)$
(We have $i^2 = -1$.)

$$= \cos(k\theta)\cos(\theta) - \sin(k\theta)\sin(\theta) + i(\cos(k\theta)\sin(\theta) + \sin(k\theta)\cos(\theta))$$

$= \cos(k\theta + \theta) + i\sin(k\theta + \theta)$ (deducted from the trigonometry rules)

$$= \cos((k+1)\theta) + i\sin((k+1)\theta).$$

Thus, for $n = k + 1$, we have $(\cos(\theta) + i\sin(\theta))^{k+1} = \cos((k+1)\theta) + i\sin((k+1)\theta)$, as expected.

As the theorem is true for $n = 1$ and $n = k + 1$, it is true for all $n \geq 1$.

Notethatinde Moivre'stheorem,thecomplexnumberisintheform $z = r(\cos\theta + i\sin\theta)$. For complex numbers in the general form $z = a + bi$, it may be necessary to first compute the absolute value and argument to convert z to the form $r(\cos\theta + i\sin\theta)$ before applying de Moivre's theorem.

Raising to a Power: Basic

Example:

Evaluate $(1-i)^6$.

In order to express $z = 1 - i$ in the form $r(\cos\theta + i\sin\theta)$, we calculate the absolute value r and argument θ as follows:

$$\text{Absolute value}: r = \sqrt{1^2 + (-1)^2} = \sqrt{2}$$

$$\text{Argument}: \theta = \arctan\frac{-1}{1} = -\frac{\pi}{4}.$$

Now, applying DeMoivre's theorem, we obtain:

$$\begin{aligned}
z^6 &= \left[\sqrt{2}\left(\cos\left(-\frac{\pi}{4}\right) + i\sin\left(-\frac{\pi}{4}\right)\right)\right]^5 \\
&= \sqrt{2}^6\left[\cos\left(-\frac{6\pi}{4}\right) + i\sin\left(-\frac{6\pi}{4}\right)\right] \\
&= 2^3\left[\cos\left(-\frac{3\pi}{2}\right) + i\sin\left(-\frac{3\pi}{2}\right)\right] \\
&= 8(0 + 1i) \\
&= 8i.
\end{aligned}$$

Example:

Evaluate $\left(\frac{\sqrt{2}}{2} + \frac{\sqrt{2}}{2}i\right)^{1000}$.

n order to express $z = \left(\frac{\sqrt{2}}{2} + \frac{\sqrt{2}}{2}i\right)$ in the form $r(\cos\theta + i\sin\theta)$, we calculate the absolute value r and argument as follows:

$$\text{Absolute value}: r=\sqrt{\left(\frac{\sqrt{2}}{2}\right)^2+\left(\frac{\sqrt{2}}{2}\right)^2}=1$$

$$\text{Argument}: \theta=\arctan 1=\frac{\pi}{4}.$$

Now, applying DeMoivre's theorem, we obtain:

$$\begin{aligned} z^{1000} &= \left(\cos\left(\frac{\pi}{4}\right)+i\sin\left(\frac{\pi}{4}\right)\right)^{1000} \\ &= \cos\left(\frac{1000\pi}{4}\right)+i\sin\left(\frac{1000\pi}{4}\right) \\ &= \cos 250\pi + i\sin 250\pi \\ &= \cos(0+125\times 2\pi)+i\sin(0+125\times 2\pi) \\ &= 1. \end{aligned}$$

Example:

Evaluate $\left(1+\sqrt{3}\,i\right)^{2013}$.

In order to express $z=1+\sqrt{3}\,i$ in the form $r(\cos\theta+i\sin\theta)$, we calculate the absolute value r and argument θ as follows:

$$\text{Absolute value}: r=\sqrt{1^2+\left(\sqrt{3}\right)^2}=\sqrt{4}=2$$

$$\text{Argument}: \theta=\arctan\frac{\sqrt{3}}{1}=\frac{\pi}{3}.$$

Now, applying DeMoivre's theorem, we obtain:

$$\begin{aligned} z^{2013} &= \left(2\left(\cos\frac{\pi}{3}+i\sin\frac{\pi}{3}\right)\right)^{2013} \\ &= 2^{2013}\left(\cos\pi\frac{2013\pi}{3}+i\sin\frac{2013\pi}{3}\right) \\ &= 2^{2013}(-1+0i) \\ &= -2^{2013}. \end{aligned}$$

Raising to a Power: Intermediate

De Moivre's Theorem

For any complex number x and any integer n.

$$(\cos x + i\sin x)^n = \cos(nx) + i\sin(nx).$$

Proof:

We prove this formula by induction on n and by applying the trigonometric sum and product formulas. We first consider the non-negative integers. The base case $n = 0$ is clearly true. For the induction step, observe that:

$$\begin{aligned}(\cos x + i\sin x)^{k+1} &= (\cos x + i\sin x)^k \times (\cos x + i\sin x)\\ &= (\cos(kx) + i\sin(kx))(\cos x + i\sin x)\\ &= \cos(kx)\cos x - \sin(kx)\sin x + 1(\sin(kx)\cos x + \cos x + \cos(kx)\sin x)\\ &= \cos[k+1)x] + i\sin[(k+1)x].\end{aligned}$$

Note that the proof above is only valid for integers n. There is a more general version, in which n is allowed to be a complex number. In this case, the left-hand side is a multi-valued function, and the right-hand side is one of its possible values.

Euler's formula for complex numbers states that if z is a complex number with absolute value r_z and argument θ_z, then:

$$z = r_z e^{i\theta_z}.$$

The proof of this is best approached using the (Maclaurin) power series expansion and is left to the interested reader. With this, we have another proof of De Moivre's theorem that directly follows from the multiplication of complex numbers in polar form.

Example:

Show that $\cos(5\theta) = \cos^5\theta - 10\cos^3\theta\sin^2\theta + 5\cos\theta\sin^4\theta.$

Applying De Moivre's theorem for $n = 5$, we have:

$$\cos(5\theta) + i\sin(5\theta) = (\cos\theta + i\sin\theta)^5.$$

Expand the RHS using the binomial theorem and compare real parts to obtain:

$$\cos(5\theta) = \cos^5\theta - 10\cos^3\theta\sin^2\theta + 5\cos\theta\sin^4\theta.$$

For an integer n, we can express $\cos(n\theta)$ solely in terms of $\cos\theta$ by using the identity $\sin^2\theta = 1 - \cos^2\theta$. This is known as the Chebyshev polynomial of the first kind.

Example:

Evaluate $\sin(0\theta)+\sin(1\theta)+\sin(2\theta)+\cdots+\sin(n\theta)$.

Applying De Moivre's formula, this is equivalent to the imaginary part of:

$$(\cos\theta+i\sin\theta)^0+(\cos\theta+i\sin\theta)^1+(\cos\theta+i\sin\theta)^2+\cdots+(\cos\theta+i\sin\theta)$$

Interpreting this as a geometric progression, the sum is:

$$\frac{(\cos\theta+i\sin\theta)^{n+1}-1}{(\cos\theta+i\sin\theta)-1}$$

as long as the ratio is not 1, which means $\theta\neq 2k\pi$. (Note that in this case, we get that each term $\sin(k\theta)$ is 0, and hence the sum is 0).

Converting this to polar form, we obtain:

$$\frac{e^{i(n+1)\theta}-1}{e^{i\theta}-1}=\frac{e^{i\left(\frac{n+1}{2}\right)\theta}}{e^{i\frac{1}{2}\theta}}\times\frac{e^{i\left(\frac{n+1}{2}\right)\theta}-e^{-i\left(\frac{n+1}{2}\right)\theta}}{e^{i\frac{1}{2}\theta}-e^{-i\frac{1}{2}\theta}}=e^{i\frac{n}{2}\theta}\frac{2i\sin\left[\left(\frac{n+1}{2}\right)\theta\right]}{2i\sin\left(\frac{1}{2}\theta\right)}.$$

Taking imaginary parts, we obtain:

$$\frac{\sin\left(\frac{n}{2}\theta\right)\sin\left(\frac{n+1}{2}\theta\right)}{\sin\left(\frac{1}{2}\theta\right)}.$$

Roots

The nth roots of unity are the complex solutions to the equation:

$$z^n=1.$$

Suppose complex number $z=a+bi$ is a solution to this equation, and consider the polar representation $z=re^{i\theta}$, where $r=\sqrt{a^2+b^2}$ and $\tan\theta=\frac{b}{a},0\leq\theta<2\pi$.

Then, by De Moivre's theorem, we have:

$$1=z^n=(re^{i\theta})^n=r^n(\cos\theta+i\sin\theta)^n=r^n(\cos n\theta+i\sin n\theta).$$

bserve that this gives n complex n^{th} roots of unity, as we know from the fundamental theorem of algebra. Since all of the complex roots of unity have absolute value 1, these points all lie on the unit circle. Furthermore, since the angle between any two consecutive roots is $\frac{2\pi}{n}$, the complex roots of unity are evenly spaced around the unit circle.

Example:

What are the complex solutions to the equation $z = \sqrt[3]{1}$?

Cubing both sides gives $z^3 = 1$, implying z is a 3rd root of unity. By the above, the 3rd roots of unity are:

$$e^{\frac{2k\pi}{3}i} = \cos\left(\frac{2k\pi}{3}\right) + i\sin\left(\frac{2k\pi}{3}\right) \text{for} k = 0,1,2.$$

This gives the roots of unity $1, e^{\frac{2\pi}{3}i}, e^{\frac{4\pi}{3}i}$, or:

$$1, \ -\frac{1}{2}+\frac{\sqrt{3}}{2}i, \ -\frac{1}{2}-\frac{\sqrt{3}}{2}i.$$

Another way to solve this equation would be to factorize $z^3 - 1 = (z-1)(z^2 + z + 1)$.

Then the solutions are $z = 1$ and the solutions to the quadratic equation $z^2 + z + 1 = 0$, which can be found using the quadratic formula.

Example:

Given positive integer n, let $\zeta = e^{\frac{2k\pi}{n}i}$ for some $k = 1,2,\ldots,n-1$ i.e., ζ is one of the nth root of unity that is not equal to 1. Show that:

$$1+\zeta+\zeta^2+\cdots+\zeta^{n-1} = 0.$$

Since ζ is an nth root of unity, we have $\zeta^n = 1$. Then:

$$0 = 1-\zeta^n = (1-\zeta)(1+\zeta+\zeta^2+\cdots+\zeta^{n-1}).$$

Since $\zeta \neq 1$, we have $1+\zeta+\zeta^2+\cdots+\zeta^{n-1} = 0$.

References

- Pythagorean-identities: brilliant.org, Retrieved 14 April, 2019
- TrigonometricAdditionFormulas: mathworld.wolfram.com, Retrieved 26 August, 2019
- Tangent, Double-Angle-Formulas: proofwiki.org, Retrieved 23 June, 2019
- Half-AngleFormulas: mathworld.wolfram.com, Retrieved 17 February, 2019
- Triple-angle-identities: brilliant.org, Retrieved 16 May, 2019
- De-moivres-theorem: brilliant.org, Retrieved 02 January, 2019

Permissions

All chapters in this book are published with permission under the Creative Commons Attribution Share Alike License or equivalent. Every chapter published in this book has been scrutinized by our experts. Their significance has been extensively debated. The topics covered herein carry significant information for a comprehensive understanding. They may even be implemented as practical applications or may be referred to as a beginning point for further studies.

We would like to thank the editorial team for lending their expertise to make the book truly unique. They have played a crucial role in the development of this book. Without their invaluable contributions this book wouldn't have been possible. They have made vital efforts to compile up to date information on the varied aspects of this subject to make this book a valuable addition to the collection of many professionals and students.

This book was conceptualized with the vision of imparting up-to-date and integrated information in this field. To ensure the same, a matchless editorial board was set up. Every individual on the board went through rigorous rounds of assessment to prove their worth. After which they invested a large part of their time researching and compiling the most relevant data for our readers.

The editorial board has been involved in producing this book since its inception. They have spent rigorous hours researching and exploring the diverse topics which have resulted in the successful publishing of this book. They have passed on their knowledge of decades through this book. To expedite this challenging task, the publisher supported the team at every step. A small team of assistant editors was also appointed to further simplify the editing procedure and attain best results for the readers.

Apart from the editorial board, the designing team has also invested a significant amount of their time in understanding the subject and creating the most relevant covers. They scrutinized every image to scout for the most suitable representation of the subject and create an appropriate cover for the book.

The publishing team has been an ardent support to the editorial, designing and production team. Their endless efforts to recruit the best for this project, has resulted in the accomplishment of this book. They are a veteran in the field of academics and their pool of knowledge is as vast as their experience in printing. Their expertise and guidance has proved useful at every step. Their uncompromising quality standards have made this book an exceptional effort. Their encouragement from time to time has been an inspiration for everyone.

The publisher and the editorial board hope that this book will prove to be a valuable piece of knowledge for students, practitioners and scholars across the globe.

Index

www.ingramcontent.com/pod-product-compliance
Lightning Source LLC
LaVergne TN
LVHW080607210926
848485LV00003B/117
* 9 7 8 1 6 4 7 2 8 3 4 9 0 *